Urban Ecosystems

With over half of the global human population living in urban regions, urban ecosystems may now represent the contemporary and future human environment. Consisting of green space and the built environment, they harbour a wide range of species, yet are not well understood.

This book aims to review what is currently known about urban ecosystems in a short and approachable text that will serve as a key resource for teaching and learning related to the urban environment. It covers both physical and biotic components of urban ecosystems, key ecological processes, and the management of ecological resources including biodiversity conservation. All chapters incorporate case studies, boxes and questions for stimulating discussions in the learning environment.

Robert Francis is Senior Lecturer in Ecology at King's College London, UK.

Michael Chadwick is a Lecturer in Freshwater & Estuarine Ecology at King's College London, UK.

Urban Ecosystems

Understanding the human environment

Robert A. Francis and
Michael A. Chadwick

LONDON AND NEW YORK

This first edition published 2013
by Routledge
2 Park Square, Milton Park, Abingdon, Oxon OX14 4RN

Simultaneously published in the USA and Canada
by Routledge
711 Third Avenue, New York, NY 10017

Routledge is an imprint of the Taylor & Francis Group, an informa business

British Library Cataloguing in Publication Data
A catalogue record for this book is available from the British Library

Library of Congress Cataloging-in-Publication Data
Francis, Robert A.
Urban ecosystems: understanding the human environment / Robert A. Francis
and Michael A. Chadwick.—1st ed.
p. cm.
Includes bibliographical references and index.
1. Urban ecology (Sociology) I. Chadwick, Michael A. II. Title.
HT241.F73 2013
307.76—dc23
2012036654

ISBN13: 978-0-415-69795-8 (hbk)
ISBN13: 978-0-415-69803-0 (pbk)
ISBN13: 978-0-203-13364-4 (ebk)

Typeset in Sabon and Gill Sans
by Book Now Ltd, London

MIX
Paper from
responsible sources
FSC® C013056
www.fsc.org

Printed and bound in Great Britain by
TJ International Ltd, Padstow, Cornwall

Contents

Illustrations

Boxes

Figures

Tables

Preface

Upon hearing that we were writing a textbook on urban ecosystems, one of our colleagues (who shall remain nameless) quipped 'It will be a very short one, then!'. This book is quite short, but not because there is little to say – far from it. The study of urban ecosystems and the field of urban ecology are growing at a fast rate, and it is difficult to keep up with the amount of published research emerging from around the world. The time of urban ecology has come, stimulated in part by a growing recognition of the importance of urban ecosystems as our most common contemporary human environment. This text aims to provide an overview of the key topics relevant to urban ecosystems and urban ecology for readers with a general background in ecology, geography or the environmental sciences. Chapters 1–5 introduce urban ecosystems and urban ecology, before moving onto considerations of urban form and structure, the main characteristics of urban region ecosystems, and then further investigation of both green space and the built environment. Following this overview, further attention is given specifically to trends and characteristics of urban species, urban nature conservation, the incorporation of ecology into urban planning and design, and the book concludes with a brief consideration of what the future might hold for urban ecosystems and ecology.

An introductory book such as this cannot be comprehensive, and despite synthesising material from almost 600 sources, much had to be left out or covered superficially. We apologise to all those authors whose research could not be included in the book, due to space limitations. It is our hope that this text will provide a starting point for further exploration of the wider urban ecology literature, as well as some of the excellent more specialised books on the topic.

Some acknowledgements are of course in order. We are grateful for permissions granted for use of figures and tables from The American Association for the Advancement of Science, Cambridge University Press, Elsevier, NASA, Springer, Taylor and Francis, and individuals acknowledged throughout for specific photographs. Tim Hardwick and Ashley Irons at Earthscan have been very helpful throughout the project, and deserve particular thanks. Rob Francis would like to thank his family for their continued patience during the writing of the book. Michael Chadwick would like to thank his family for their understanding and support. Much of the book is based on subjects taught by the authors within undergraduate and postgraduate programmes in the Department of Geography at King's College London, which have been shaped by both colleagues and students over the last six years. We thank them all for their comments and insight, particularly graduates of the MSc Sustainable Cities programme.

Rob Francis and Michael Chadwick
London, August 2012

Chapter 1

An introduction to urban ecology and urban ecosystems

1.1 About this book

The purpose of this book is to provide a broad overview of urban ecology and urban ecosystems for students, researchers and practitioners. Research into urban ecology has blossomed in the last two decades, bringing with it both increased interest and understanding of urban ecosystems, ecological processes and biodiversity. Although this text cannot provide exhaustive coverage of all aspects of what is a broad and interdisciplinary area of study, it will equip the reader with a sound understanding of the main concepts and topics relevant to the field, and act as a primer for further exploration of the literature.

The focus of the text is on the 'urban region' (defined and discussed in more detail below), and includes many forms of urban ecosystems. As Pickett *et al.* (2001) note in their key paper on linking ecological, physical, and socio-economic components within urban areas, an ecological understanding of an urban region must include the less densely populated suburbs, satellite towns and villages, and rural/urban gradients because species, materials and energy flow between these different components. This book therefore endeavours to cover a wide spectrum of urban environments to fully equip the reader with a suitably broad understanding.

This text does maintain something of a focus on heavily urbanised areas, however, as such ecosystems are the most extreme modifications that humans have made to the natural environment, and represent perhaps the most interesting systems for studying the effects that urbanisation and human activity can have on ecological patterns and processes. They are also the most exciting opportunities we have for environmental improvements and monitoring future broad-scale ecological changes. Urban regions are, from an ecological point of view, huge and prolonged experiments that have changed (to a greater or lesser extent) the characteristics found in natural ecosystems so that something new, strange and wonderful has emerged. This book aims to summarise and elaborate on some of these aspects of urban regions.

The book is divided into nine chapters, as follows:

 Chapter 1 provides an overview of what urban ecology is, what urban ecosystems are, and how some key urban terms have been defined. It also briefly summarises the history of urbanisation and reflects on why urban ecosystems are important or interesting ecologically.

- Chapter 2 considers the form, structure and dynamics of urban environments at both regional and landscape scales, highlighting some ways in which spatial organisation and temporal dynamics may relate to ecological patterns and processes. It also considers the three-dimensional structure of urban environments. These topics are important for developing a full understanding of the biophysical template urban areas represent, which both shapes and is shaped by ecological processes.
- Chapter 3 gives a broad overview of the urban region as an ecosystem, highlighting some of the key characteristics associated with urban environments.
- Chapter 4 examines urban ecosystems found within urban regions, focusing on those that may be classified as 'green space'. This includes parks and recreational spaces, gardens, lawns, allotments, brownfield and wasteland sites, remnant woodland, and urban rivers and lakes.
- Chapter 5 looks at urban ecosystems found within urban regions that are formed in large part from the built environment, including built surfaces, living roofs and walls, terrestrial and aquatic infrastructure such as roads, pavements, railways, underground rapid transit systems, sewerage systems and canals, and infrastructural trees.
- Chapter 6 covers urban species, discussing common trends observed in the types of species and assemblages found in urban ecosystems, including seral changes and ways in which species respond to urban environments.
- Chapter 7 considers nature conservation in urban regions, looking at the preservation and restoration of species and green space within an urban context, as well as the possibilities represented by urban reconciliation ecology. It also reflects on conservation governance and the growing role of citizen science.
- Chapter 8 briefly examines how ecology has been and continues to be incorporated into urban planning and design, looking at ecologically sensitive urbanisation, the planning and design of green space and green networks, and examples of ecological and environmental engineering in urban regions.
- Chapter 9 summarises some likely future trends relating to urban ecosystems, including urban growth, biodiversity, sustainability, and planning, and finally draws out some key areas for future research in urban ecology.

Each chapter incorporates a brief summary and some open questions that may be used as starting points for wider class discussions or further individual investigation.

1.2 Urban ecology: what it is and is not

Ecology as a discipline developed from a background of natural history in the nineteenth century; the term (*Oekologie*) was first used by Haeckel (1866), though the formalisation of the scientific discipline from the more descriptive precursor of natural history really occurred in the early twentieth century with the work of ecologists such as Frederic Clements, Henry Gleason and Arthur Tansley. 'Ecology' may refer to the interactions and relationships between organisms and the abiotic environment, or the science of studying these. It is usually, as in the case of urban ecology, further refined within the context of a particular ecosystem or environment (tropical ecology, desert ecology), groups of organisms or their defining traits (mammal ecology, invasive species ecology) or space or timescale (landscape ecology, palaeoecology). Importantly, ecology must incorporate the living world (and in particular the nonhuman living world). In many ways it is the oldest science, as humans have always had cause to investigate interactions between animals, plants and their environment, to make them successful hunter-gatherers and eventually agrarian species. The formalisation of the discipline has been surprisingly recent, but the discipline and its various subdisciplines are now major parts of the scientific canon.

The discipline of urban ecology (following the definition of ecology outlined above) has emerged relatively recently. The term 'urban ecology' was originally (in the early part of the twentieth century) used by human geographers and sociologists who applied ecological terms and concepts to explain human influences on spatial patterns and processes within cities (Braun, 2005). This was in essence the 'human ecology' of urban regions rather than urban ecology as we know it now; the environment and 'nature' of, and within, urban regions was considered in relation to human activities and social patterns, but it was not a focus of the discipline. Of course, the sorts of scholarly activities that would eventually feed into the discipline of urban ecology that is the focus of this book have been occurring for centuries; studies (mainly botanical) of both spontaneous and cultivated urban species and assemblages (particularly wall and garden flora), have been recorded in some urban environments since the 1600s (Sukopp, 2002), for example. The 'new' urban ecology, incorporating urban areas more holistically and within 'ecosystem' concepts, did not really develop until the 1970s.

After this point, there is an increase in studies that focus more specifically on the ecology of urban areas (and a concomitant decline in the 'human' urban ecology) in literature databases such as ISI's Web of Knowledge (Figure 1.1). These early studies were mainly case studies or surveys, but were paving the way for more considered and holistic investigations. Urban ecology remains something of a minority discipline but an important one, and the questions that urban ecologists seek to answer are crucial to societies around the world. The rise in urban environments as a focus for both theoretical and applied investigation can be seen in, for example, the establishment of The United Nations Human Settlement Programme in 1978, which is specifically focused on urban regions; the establishment of academic journals such as *Landscape and Urban Planning* (1974) and *Urban Ecosystems* (1997); the emergence of university and research centres focusing on urban environments; and, at a more local scale the establishment of organisations like 'Urban Ecology' in Washington in 1975 and the 'Trust for Urban Ecology' in London in 1976. The science of ecology is an important aspect of all of these urban endeavours. Importantly, urban ecologists do not seek solely to apply ecological theory to an understanding of urban environments, but also to use urban environments to develop an understanding of fundamental ecological theory.

1.3 What are urban ecosystems?

The term 'ecosystem' is used in a very broad way to refer to spatially defined areas of the natural environment. The breadth of interpretation of this term can result in some confusion. When conceptualising an urban region for example, is the entire region the ecosystem, or is it a park within the region, or a tree within the park? The simple

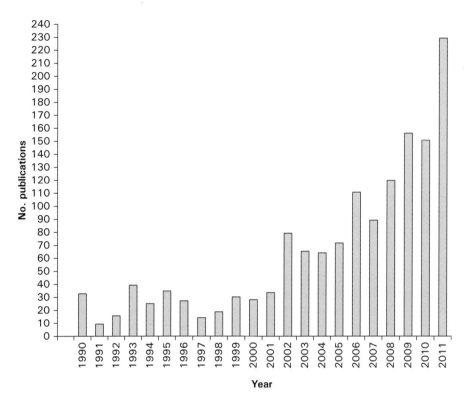

Figure 1.1 Number of papers published each year found using a search of 'urban ecosystem*' or 'urban ecology' in ISI's Web of Knowledge, published between 1990 and 2011. A total of 1496 papers were found using the search.

answer is that they may all be considered ecosystems. Usually, ecosystems are broadly defined as 'a biotic community or assemblage and its associated physical environment in a specific place' (Pickett and Cadenasso, 2002, p. 2; see this paper for an elegant discussion of the ecosystem concept). As Pickett and Cadenasso (2002) note, this definition (a refinement of Tansley's (1935) original and enduring definition) is scale independent. This means that a rotting orange on the ground may be considered as much an ecosystem as a river or patch of woodland. However, broad spatial scales are usually the focus for ecosystem science. Sometimes, a fine-scale ecosystem is termed a 'biotope'. Use of this term usually assumes homogeneity of environment and the presence of a particular ecological community, and thus does not often reflect the heterogeneity of conditions typically found within ecosystems. Consequently, we use the term 'ecosystem' across all spatial scales in this book.

Essentially, ecosystems are spatially nested and are defined relative to the specific ecological patterns or processes that are being considered (Rebele, 1994; Pickett *et al.*, 2001). Urban ecosystems include the habitats that make up urban areas, such as buildings, walls, roads, parks, gardens, rivers, canals, and so on, the species that live in them, and the assemblages that species form.

We use the term 'assemblage' rather than 'community' in this text to designate two, or more populations of species coexisting within a defined area (i.e. a discrete spatial unit), because of the potential conflation with the use of 'community' as the formal classification of particular species associations, such as a typical old-growth woodland community for a given biogeographical region. Although urban 'communities' have been explored and classified (see, e.g. Rodwell *et al.*, 2000), given the wide range of species associations found in urban environments, we find the term generally unhelpful and so avoid it here.

1.4 Defining 'urban'

'Urban' can be a difficult term to pin down, despite the way in which it is defined and used having significant implications for urban ecology. In part, this is because the term seems intuitive – most people when asked would be able to give an indication of what 'urban' is with some confidence, and at a superficial level drawing a distinction between 'urban' and 'rural' (for example) seems straightforward. But what is 'urban', technically? Definitions of urban areas generally vary between designations based on population density, administrative boundaries and/or spatial dominance of the built environment, and may differ substantially between what an academic would describe as 'urban' in a scientific paper, and how a governmental body would define the term operationally. The United Nations Statistics Division (2011) notes that 'urban' is 'not amenable to a single definition applicable to all countries' and that definitions 'based solely on the size of the population of localities do not always offer a satisfactory basis for classification, especially in highly industrialized countries' (p. 1). Consequently, the UN summarises many operational definitions of urban for differing countries (see Table 1.1). Although the majority of definitions are based on population (either in absolute terms or as population density), some countries more explicitly include the density of the built environment: in Norway for example, an urban settlement is considered to be a collection of buildings with less than 50 m between individual buildings, and inhabited by at least 200 people (Statistics Norway, 2011). Table 1.2 highlights the relative balance between definitions based on population, built environment and administration.

These differences can be important, particularly if planning or management decisions are being made in relation to 'urban' areas. In particular these definitions do not always translate into heavily developed, primarily artificial, built ecosystems that people may typically associate with urban environments. For example, in some highly developed regions, urbanisation has become largely extensive rather than intensive, with relatively few people living in less densely arranged buildings and highly developed infrastructure (e.g. Kasanko *et al.*, 2006). Are the ecological differences between these areas and heavily developed ones sufficient for a 'non-urban' designation? The answer is frequently complex, and site-specific. Furthermore, the population density that constitutes 'urban' for one country may be far less than that for another (Albania for example classifies urban as those areas with more than 400 residents, while Japan requires a population of over 50,000; see Table 1.1). Direct comparisons can therefore be difficult.

On top of this, there are several synonyms that are used for 'urban' (such as city, town, metropolis, settlement), as well as subcategories that are frequently referred to

Table 1.1 Definitions of 'urban' in selected countries around the world

Country	Definition of 'urban'
Africa:	
Algeria	Groupings of 100(+) constructions spaced less than 200 m from one another
Ethiopia	Localities of 2000(+) inhabitants
Kenya	Areas of 2000(+) inhabitants with associated infrastructure
Lesotho	Administrative headquarters and areas of rapid growth
Senegal	Agglomerations of 10,000(+) inhabitants
South Africa	Areas with some form of local authority
Uganda	Gazettes, cities, municipalities and towns
Zambia	Localities of 5000(+) inhabitants, most of whom are not engaged in agriculture
Asia:	
Cambodia	Commune areas with total population 2000(+), 200(+) people km^{-2}, and less than 50% of males employed in agriculture
China	Cities designated by the State Council. Further classifications depend on whether areas have 'district' status, the location of local government, and the population density of the settlement
India	Towns designated according to administrative capacity; or having 5000(+) inhabitants, 400(+) people km^{-2} and at least 75% of the male population employed in an industry other than agriculture
Iran	A district with a municipality
Israel	Localities with 2000(+) residents
Japan	City with 50,000(+) inhabitants, with 60%(+) of houses in a built-up area and 60%(+) of the population engaged in manufacturing, trade or service industries
Republic of Korea	Places with 50,000(+) inhabitants
Turkey	Province and district centres
Europe:	
Albania	Towns and industrial areas of 400(+) inhabitants
Czech Republic	Localities of 2000(+) inhabitants
France	Communes of 2000(+) residents with no more than 200 m between houses
Iceland	Localities of 200(+) inhabitants
Netherlands	Municipalities with 2000(+) inhabitants
Poland	Settlements of an 'urban' type
Spain	Localities of 2000(+) residents
UK	Settlements of 10,000(+) inhabitants
North America:	
Canada	Places of 1000(+) inhabitants, with a population density of 400(+) km^{-2}

(Continued)

Table 1.1 (Continued)

Country	Definition of 'urban'
Cuba	Towns with an administrative function or 2000(+) inhabitants
Greenland	Localities of 200(+) inhabitants
Haiti	Administrative centres of communes
Jamaica	Localities of 2000(+) inhabitants, with 'urban' characteristics
Mexico	Localities of 2500(+) inhabitants
Panama	Localities of 1500(+) inhabitants with essential infrastructure
United States of America	Agglomerations of 2500(+) inhabitants, with population densities of 386(+) km^{-2}. Further classified into 'urbanised areas' (50,000(+) inhabitants) and 'urban clusters' (2500–50,000 inhabitants)
Oceania:	
Australia	'Urban centre' is a cluster of 1000(+) inhabitants
New Zealand	All cities, towns and districts with 1000(+) inhabitants
South America:	
Argentina	Populated centres with 2000(+) inhabitants
Brazil	Administrative centres of municipalities and districts
Columbia	Administrative centres of municipalities
Ecuador	Capitals of provinces and cantons
Paraguay	Cities, towns and administrative centres
Peru	Populated centres with 100(+) residences
Uruguay	Cities
Venezuela (Bolivarian Republic of)	Centres with 1000(+) inhabitants

Definitions are summarised from the *United Nations Demographic Yearbook 2009–10*, and are based on the definition used for the national census for each country (see United Nations Statistical Division, 2010, for further details and a full list of countries).

Table 1.2 The number of definitions of 'urban' in the *United Nations Demographic Yearbook 2009–10* that are based on some measurement of population, built environment or infrastructure and/or administration

	Population	Built environment or infrastructure	Administration
Africa	9	7	14
Asia	20	15	20
Europe	23	10	9
North America	11	5	7
Oceania	5	0	2
South America	5	3	6
Total:	73	40	58

Note
Definitions that use more than one measurement or indicator are counted more than once.

(e.g. suburban, peri-urban, ex-urban, urban core, urban fringe, satellite and periphery, amongst others; see MacGregor-Fors, 2011). Clearly some level of universal agreement and standardisation of urban terminology is required, at least with respect to ecology; this issue has been tackled by both McIntyre (2000) and MacGregor-Fors (2011). Perhaps the most useful definition is that offered by MacGregor-Fors (2011, pp. 347–348), which classifies 'urban' as:

> populated areas provided with basic services (e.g. homesteads, electricity and water supply, drainage), where more than 1000 people km^{-2} (>10 inhabitants ha^{-1}) live or work, and an important proportion of the land (>50 per cent), in a "city-scale" [*i.e. a spatially delineated area such as a city or town, our clarification*], is covered by impervious surfaces (e.g. buildings, streets, roads).

In contrast, anything that does not meet this definition may be simply regarded as 'non-urban'. This is not a foolproof definition of course, and may not have direct ecological relevance – a large industrial estate on the edge of an 'urban' region may for example not support a high population density (living or working) but the land surface is still highly developed and may function ecologically in the same or a similar way as a city centre. To overcome this, MacGregor-Fors (2011) also recommends that a measure of urban intensity be recorded when discussing urban ecology, based on the land cover of impervious surfaces. He suggests a classification where sparsely developed urban areas have 0–33 per cent built cover, moderately developed urban areas have 34–66 per cent built cover, and highly developed urban areas have 67–100 per cent built cover. It is, however, important to remember that such definitions hang on the spatial scale being considered for the ecological pattern or process under examination.

To clarify: in the 'urban' definition given above, an 'urban' area requires that the settlement under consideration (the 'city-scale') has at least 50 per cent built cover over the entire area, and within this there may be areas of sparse or moderate development that are part of the heterogeneous urban landscape, with less than 50 per cent cover for a given unit (parks, gardens, rivers, etc.). At the 'settlement-' or 'city-scale', such areas are considered urban, though the effects of the broader urban environment may (or indeed may not) be less relevant at finer spatial scales. For example, although the biodiversity of particular species or species groups may be suppressed overall within an urban region, locally (such as in some in gardens or parks) species richness may be higher than in a more 'natural' ecosystem (McKinney, 2008).

One advantage of classifying areas based on the percentage cover of the built environment is that it is a form of much-needed standardisation (rather than subjective decisions on what is 'suburban' or the 'urban centre' for example), and it does remove the need for the use of confusing terms such as 'suburban, ex-urban, peri-urban' etc., which can be replaced by quantified 'sparsely developed' or 'moderately developed' areas within the urban region. It does, however, raise the broader question of how the settlement itself is delineated – are 'sparsely developed' areas on the fringe of the settlement to be included as part of the 'urban region', or are they non-urban, on account of their <50 per cent built surface cover, or less dense population?

The use of subcategories to define gradients of urbanisation can also be problematic. Broadly, this is usually conceptualised (starting in the centre of an urban area and working out) as the 'centre', 'peri-centre', 'fringe', 'suburbs' and finally 'rural'.

However, as MacGregor-Fors (2011) notes, drawing distinctions between 'centre and 'peri-centre' and 'peri-centre, fringe and suburbs' can be difficult and is often highly subjective. It is also not always very ecologically meaningful. It is important, however, to note 'exurban' areas, which are those that may be influenced ecologically by an urban area (e.g. by urban climate or hydrologic effects) but which are not (for whatever reason) included with the urban area itself – these are therefore influenced by urbanisation but are not necessarily 'urban'.

What may be learned from this is that the term 'urban' is not always well defined and varies between individual studies, regions and countries. The definition provided by MacGregor-Fors (2011) is perhaps the most useful for standardising how the term is used, but still has problems of subjectivity in how the 'city-scale' is delineated (and which may end up simply based on administrative boundaries). Rather than using 'planning' terminology such as 'suburbs' or 'fringe', a quantification of built land cover to use as a basis for classification is the simplest way of standardising levels of 'urban'. Whether these classifications are ecologically meaningful remains to be determined. The key lesson is that for all reported investigations, the urban context should be clearly expressed to allow comparison between studies – the definitions given above provide a useful starting point for this. As this book aims to provide an overview of urban ecosystems and ecology, we interpret 'urban' broadly as a combination of (1) a high proportion of built environment and (2) a relatively high population density within a regional context. Within this text, we adopt a simple approach to drawing spatial distinctions within urban regions. A contiguous area of the built environment, which may contain within it patches of green space or other typically 'non-urban' ecosystems, is referred to as the 'urban complex'. (This is essentially what Forman (2008) and some authors refer to as the 'metropolitan area', within which is contained the urban centre or 'city' – we avoid these terms as 'metropolitan' is used in different ways within the urban ecology literature, and 'city' is an unhelpful term ecologically – see Section 1.5.) The urban complex is positioned within an 'urban region' which, following Forman (2008) represents the urban complex and the surrounding exurban land, which is directly influenced by the urban complex (e.g. via local climatic effects, pollution, etc.).

These are of course generalisations, and quantifying the extent of any specific urban region would require detailed environmental measurements. Nevertheless, the broad distinction between the urban complex and wider urban region is easy to conceptualise. Where appropriate, distinctions between the centre and edge of the complex/region are made, but we avoid the pseudo-detail of 'periphery', 'suburb', 'fringe', etc. Anything outside of the urban region is simply considered 'non-urban' for the purposes of the book.

One confounding action frequently taken in urban ecology (and urban studies in general) is the interchange of the term 'urban' and 'city'. This probably comes from the original etymology of 'urban', which stems from *urbs*, Latin for 'city', and may therefore seem technically correct. However, in its modern application, while a city may certainly be considered 'urban', what is 'urban' is not necessarily a city.

1.5 What are cities and towns?

Cities are complex and fluid entities, and the ways in which they are defined reflects the nature of the academic discipline framing the question. In his major text on urban landscape ecology, *Urban Regions: Ecology and Planning Beyond the City*, Forman

(2008), mainly concerned with spatial designations, defines them simply as 'a relatively large or important municipality' within the broader urban region (p. 6). This therefore places cities as large and heavily developed urban complexes (with high populations and a dense built environment). Davison and Ridder (2006), focusing on society–nature relationships, describe them as 'sustained events' that incorporate humans and non-humans, ecology and technology, and real and virtual realities (p. 307). Clearly, the definition of a city is very elastic and context-dependent.

The etymology of 'city' is complex and the ways in which we define cities means that there is little merit in painstakingly teasing out how and why the term originated and what it has been applied to previously. Just as a point of clarification it is worth noting that the original English term 'city', as applied to an urbanised location, is an adaptation of the High Medieval period French *cité*, and was used following the Norman conquest of England to designate large and/or important towns. This later became modified to incorporate the presence of a cathedral in Christian Europe, as a further designation of importance, though this is now no longer the case; for example the designation of 'city' can be conferred by royal authority in the UK. In other locations it is also used to designate importance (e.g. Forman, 2008), but in others, such as North America, it may indicate governmental autonomy or a relatively high level of spatial or political organisation (compared to a 'town'). In one of the few academic studies of what constitutes a city, Gates (2003) considers that historically such urban areas were characterised by high population density, social stratification, the practice of arts and sciences, public architecture, foreign trade and some form of social community. Although all of these remain true for modern cities, these are all social and cultural indicators and, although relevant to urban ecology, are not defining characteristics in themselves.

A 'town' is generally considered to be a settlement that displays some degree of regularity and planning in its construction, and which is larger than a village. The term also relates to administration and importance of the settlement; the governing body of a town is generally larger and more powerful than that of a village, but may be less important than that of a city. In terms of size, density, population and other quantitative variables that may be used to quantify urbanisation and which may have ecological relevance, towns and cities are generally no different. A general rule of thumb would be that cities have larger and denser populations, greater spatial extent, and greater density of the built environment, along with a larger degree of planning and autonomy of government, but there is no consistent technical distinction to be drawn between the two. (city & towns)

For the purposes of this book, we consider that the terms 'city' and 'town' have no ecological relevance, other than that they generally designate an area of dense population and built environment. Though the terms may be used in the text, this will only be in an official sense, i.e. when a settlement is designated as such by its governing authorities. Otherwise, 'highly developed' is a more appropriate descriptor for such urban regions.

1.6 A short history of urbanisation

The process of urbanisation arguably began when humans first constructed shelters that allowed them to remain in one place for extended periods of time. As such developments would only have occurred where resources (particularly fresh water) were plentiful and reliable, population increase (from either immigration or reproductive success) would have followed, and the two principle components of an urban region, i.e. relatively high population densities and coverage of the built environment, were starting to materialise. Larson *et al.* (2004) suggest that the development of buildings originated from the use of cave complexes by early humans, who then expanded and replicated essential cave structures in resource-abundant locations, particularly as population pressure increased. The development of building techniques such as the manufacture of stone blocks and bricks (and mortar) from 9000 BC onwards is linked to the development of the first urban 'regions' (relatively large settlements such as Jericho and Çayönü, and early 'cities' such as Uruk and Ur). Permanent settlements are also associated with the rise of cultivation and agriculture (Larson *et al.*, 2004), creating the principle of an urban region and population centre utilising resources from the land around it, and therefore having some level of environmental influence and impact beyond the immediate urban complex (the ecological 'footprint'; see Grimm *et al.*, 2008; Section 3.2.2).

Such early 'urbanisation' of course does not much relate to the concept as we perceive it today, other than providing the initial elements. Although relatively large urban areas developed prior to the industrial revolution in various parts of the world (e.g. Constantinople [Istanbul], Paris, London, Peking [Beijing]) it is only with the development of better building materials, techniques, infrastructure, sanitation and medicine that the large and dense urban complexes that are so prevalent today have proliferated. Berry (1990) notes that around at the start of the eighteenth century, less than 10 per cent of the global population was considered 'urban', or living in settlements that could be designated as 'cities' or 'towns', though cities in particular were central to many regional and even global economies. Such urbanisation (in this case based on populations and economic significance of settlements) in the 1700s was centred mainly in modern Europe (29 urban 'centres') and China (21), with fewer large urban settlements in the Middle East (12), India (8) and East Asia (8) (see Berry, 1990). Most of these urban centres were small compared to current standards, with many having fewer than 100,000 residents and only five having populations greater than 500,000; the largest urban region of the period was Istanbul (Constantinople) with around 700,000 inhabitants.

This broad pattern of urbanisation remained until the twentieth century, though an important shift in regional patterns of urbanisation occurred with the industrial revolution, initially in the UK. Most urban growth prior to this occurred in the urban centres (usually the capital city of a country or region), but the development of industry in towns throughout the UK led to population concentration in many other urban regions, such that the proportion of urban population in England had risen from around 17 per cent in 1700 to around 27.5 per cent by 1800, most of the growth occurring outside London in the more industrial north (Berry, 1990). This was the beginning of a fundamentally urban society, wherein the majority of people live in areas that may be classified as 'urban', and was facilitated by housing and infrastructural developments that allowed greater population densities (with concomitant health impacts, at least initially) and greater transportation possibilities. The latter also led to the development of more distinct urban gradients as people worked in urban regions but resided outside of the urban 'core' (see Chapter 2). This style of

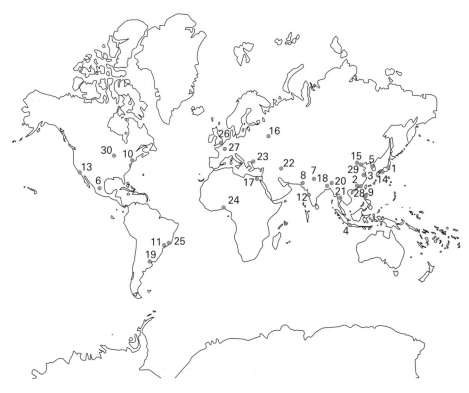

Figure 1.2 Location of the 30 largest urban regions in the world, classified according to population size (2012). 1 = Tokyo, Japan; 2 = Guangzhou, People's Republic of China; 3 = Shanghai, People's Republic of China; 4 = Jakarta, Indonesia; 5 = Seoul, South Korea; 6 = Mexico City, Mexico; 7 = New Delhi, India; 8 = Karachi, Pakistan; 9 = Manila, Philippines; 10 = New York, USA; 11 = São Paulo, Brazil; 12 = Mumbai, India; 13 = Los Angeles, USA; 14 = Osaka–Kobe–Kyoto, Japan; 15 = Beijing, People's Republic of China; 16 = Moscow, Russia; 17 = Cairo, Egypt; 18 = Kolkata, India; 19 = Buenos Aires, Argentina; 20 = Dhaka, India; 21 = Bangkok, Thailand; 22 = Tehran, Iran; 23 = Istanbul, Turkey; 24 = Lagos, Nigeria; 25 = Rio de Janeiro, Brazil; 26 = London, UK; 27 = Paris, France; 28 = Shenzhen, People's Republic of China; 29 = Tianjin, People's Republic of China; 30 = Chicago, USA. (Compiled from Thomas Brinkhoff: The Principal Agglomerations of the World, http://www.citypopulation.de, accessed 22/08/2012. Note that population estimates vary between sources based on where region boundaries are placed and quality of census or other survey data.)

urbanisation was to become very apparent globally in the following decades and centuries. Due to increasing industrialisation, urban regions have grown substantially since 1800, firstly in Europe or in European colonies, including North America. This growth mainly occurred in the twentieth century: in 1900, only 43 urban regions had populations greater than 500,000, and fewer than 10 countries had greater than 25 per cent of their population in urban centres of over 10,000 (Berry, 1990). In contrast, by 2000, many of the more developed countries had urban populations greater than 75 per cent (and all greater than 50 per cent) and only a few developing countries maintained levels less than 25 per cent. The twentieth century was also the period that saw the great urbanisation of North America and much of Asia, in many cases highly planned (unlike in many older urbanised countries such as in Europe) and again with distinct gradients from centre to edge of the urban complex, both in terms of population concentration and economy.

Currently, the largest urban centres in the world are broadly distributed (Figure 1.2), and future growth is most likely in Africa, Asia and South America, where it is predicted that urban populations may double between 2000 and 2030, such that developing countries will contain over 80 per cent of the global urban population (UNFPA, 2007, UN-Habitat, 2011). With increased pressure to contain lateral spread of urban areas in many regions to limit environmental impacts (or simply due to lack of space or resource supply), growth is particularly being made possible by architectural advances and use of the vertical dimension, as shown more recently by high-rise housing developments in major urban regions such as London, Singapore, Hong Kong, Shanghai, Beijing and Melbourne (Yuen *et al.*, 2006). In some (both developed and developing) regions, a process of counter-urbanisation has been taking place in recent decades, with a decline or retraction of urban spatial extent and a decrease in urban population densities as more people live outside urban complexes to better enjoy the less degraded environment, a trend made possible by technological and infrastructural advancements, particularly the revolution in information technology that allows more flexible working practices and frees some of the population from urban residence (though not necessarily urban employment) (Anas *et al.*, 1998). As the industrial revolution dramatically accelerated urbanisation, so the information revolution is changing the pattern of urbanisation – this is not making populations non-urban, but rather easing the population pressure of heavily urbanised regions to create urbanised societies with fewer dense urban cores (e.g. Berry, 1990). Kim (2007) demonstrates that in many major American urban regions, urban land area has increased but population growth has remained relatively stable, leading to an overall decrease in urban density.

Although they may only cover between 0.5 per cent and 3 per cent of the global land surface, depending upon remote sensing classification method (see e.g. Schneider *et al.*, 2009), urban regions have an importance that goes far beyond their spatial extent, both for human society and for the current and future environment. Some of this relates to negative impacts (see Chapters 2 and 3), but it is also important to remember the positive aspects of urban regions. Some of these are considered in the remainder of this chapter.

1.7 Why are urban regions interesting and important ecologically?

Urban regions and the ecosystems of which they are composed are interesting for several key reasons:

1   *Ecological novelty.* Urban ecosystems have emerged relatively recently in comparison to many more natural systems, and although the fundamental ecological processes that operate within them (e.g. competition, disturbance, population growth and decline) are in essence no different to those occurring in other ecosystems, the environmental conditions created in urban regions may change the

intensity, duration or frequency of such processes in relatively unique ways. For example, urban animal population densities can far exceed those that can be observed in other environments for some species (such as the common urban pest *Columba livia*, the rock dove or pigeon), often due to a change in one or more ecological processes. For the rock dove, the high abundance of food resources and nesting habitat in urban areas leads to a decline in competition between individuals, and allows for dramatic population growth.

In terms of natural selection, the sudden emergence of such urban ecosystems has presented a whole new range of challenges for urban species, and this has led to some interesting responses within urban areas (see Section 6.8). The significance of the recent emergence of urban regions for human society should also not be overlooked; the sudden 'urbanisation' of societies that have historically (>*c*.190,000 years) existed at low densities and with direct and frequent interaction with many other species and a range of physical environments, arguably led to the first 'romanticising' of the wilderness (which in this context may be loosely termed the pre-urban) and indirectly to many aspects of nature conservation, from the formation of the first nature reserves and parks, to the importation of 'wilderness' into cities (e.g. urban parks, gardening, horticulture), and the financial support of species and ecosystems far removed from those people providing the support (e.g. Rosenzweig, 2003; Hinchman and Hinchman, 2007). In particular, heavily developed and densely constructed urban regions are very new, with the earliest large urban settlements dating from the Sumerian cities of Mesopotamia (within the last 5000 years; Gates 2003), and more modern highly urbanised cities dating from the industrial revolution (after *c*.1750).

Urban regions are also somewhat unique because of the dominance of the built and social environment in urban ecosystems, which are major drivers of ecosystem patterns and processes (see Pickett and Cadenasso, 2009; Pickett and Grove, 2009; Figure 1.3). Although all extant ecosystems contain some level of anthropogenic impact (e.g. Hobbs *et al.*, 2006), their particular dominance in urban regions makes them something of a special case, and adds to their complexity at many levels, whether this is in building design, spatial arrangement of ecosystems, distribution of biota, or socio-ecological processes. Despite this complexity, they can therefore tell us about general ecological processes as well as the specifically urban, and potentially inform an understanding of how other ecosystems may respond to anthropogenic impacts and future urbanisation. As Grimm *et al.* (2008, p. 756) note, they are 'microcosms of the kinds of changes that are happening globally'.

This ecological novelty is also reflected in the types and diversity of species found within urban regions. They can exhibit new ecologies and novel species assemblages, a 'recombinant' or 'mixo-ecology' that has not been seen before and which may be particularly dynamic and changeable (see Hinchliffe and Whatmore, 2006; Section 6.4). This is due both to the destruction of pre-urban habitat (including loss of seed banks) that removes species from the area, as well as the intentional and spontaneous introduction and establishment of species from around the globe (e.g. Soulé, 1990; discussed more in Chapter 6). Urban regions certainly fit within the concept of 'novel' or 'emerging' ecosystems as coined by Hobbs *et al.* (2006), as they include new species combinations, have

[handwritten margin note: socio-ecological = coherent system of biophysical & social factors that regularly interact in a resilient, sustained manner]

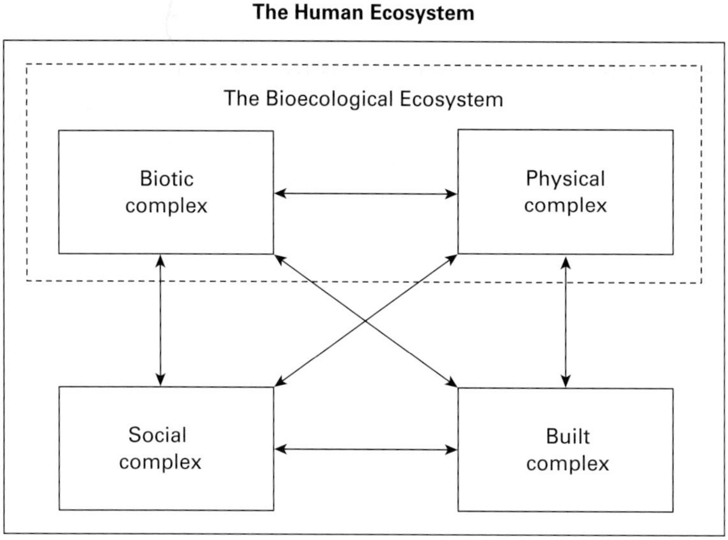

The Human Ecosystem

```
┌─────────────────────────────────────────────────────────┐
│  ┌ ─ ─ ─ ─ ─ ─ ─ ─ ─ ─ ─ ─ ─ ─ ─ ─ ─ ─ ─ ─ ─ ─ ┐        │
│          The Bioecological Ecosystem                     │
│  │                                              │        │
│     ┌──────────────┐          ┌──────────────┐          │
│  │  │   Biotic     │◄────────►│   Physical   │  │        │
│     │   complex    │          │   complex    │           │
│  │  └──────────────┘          └──────────────┘  │        │
│  └ ─ ─ ─ ─ ─ ─ ─ ─ ─ ─ ─ ─ ─ ─ ─ ─ ─ ─ ─ ─ ─ ─ ┘        │
│     ┌──────────────┐          ┌──────────────┐          │
│     │   Social     │◄────────►│    Built     │          │
│     │   complex    │          │   complex    │          │
│     └──────────────┘          └──────────────┘          │
└─────────────────────────────────────────────────────────┘
```

Figure 1.3 The 'human ecosystem' concept, illustrating some of the key components and interactions found within urban ecosystems. In contrast to more natural systems, urban ecosystems are particularly heavily influenced by the 'social' and 'built' complexes. (Reproduced from Pickett and Grove, 2009, with permission.)

relatively novel or different forms of ecosystem function, and are the result of human agency. This novelty has traditionally been viewed as something threatening and anathema to nature, but there are now calls for urban regions to be viewed as exciting and cosmopolitan regions that present interesting assemblages and that offer opportunities for ecological engineering and other forms of proactive conservation to create biodiverse urban regions (such as reconciliation ecology; see Rosenzweig, 2003; Lundholm and Richardson, 2010; Francis and Lorimer, 2011; Section 7.4). This is particularly important as the majority of the global population now lives in urban regions (UNFPA, 2007), and so most people will obtain their initial (and possibly fundamental) and most frequent interactions with 'nature' in an urban context – regardless of how 'novel' such nature may be.

2   *Extent of impacts.* Despite their modest global land coverage, urban regions can have substantial local, regional and even global environmental impacts (Grimm *et al.*, 2008). These include habitat destruction and modification, construction of artificial habitats, changes in biodiversity and species assemblages, natural selection, microclimate, hydrological dynamics, increasing pollution, soil compaction and decline in ecosystem services, among other things. These are discussed in more detail in Chapter 3.

3   *Global laboratories.* Although all urban areas (as with any other ecosystem) have their own histories and idiosyncrasies, the nature of their construction (as idealised human habitat) means that many of their environmental conditions are somewhat standardised. These include, for example, abundant resources, high productivity

and relatively stable climates that replicate, to some extent, sub-tropical environments (Shochat *et al.*'s (2006) 'pseudo-tropical bubbles'). This does not mean that all urban areas are comparable across the globe of course, but it does mean that the relative magnitude of difference between urban regions that are geographically separate is probably less for many environmental characteristics than would be the case for, e.g. forest or grassland ecosystems that are in different ecozones. This possibility does need further research and quantification, however (e.g. Francis *et al.*, 2012).

This also makes urban ecosystems experimental laboratories that have global replication, and are therefore suitable for robust scientific investigations. Currently, in-depth and interdisciplinary studies have occurred in a few isolated urban regions, such as the Long-Term Ecological Research sites in Baltimore and Phoenix in the USA (see Box 1.1), but a more global synthesis is underway.

Box 1.1 Urban Long-Term Ecological Research (LTER) sites: Baltimore and Phoenix

Some of the most useful and detailed research into urban environments in the last two decades has emerged from LTER sites in the USA: Baltimore, Maryland and Central Arizona-Phoenix. LTER sites are maintained by the National Science Foundation, and are aimed at developing detailed and long-term datasets of both ecological and socio-ecological patterns and processes, in order to 'provide the scientific community, policy makers, and society with the knowledge and predictive understanding necessary to conserve, protect, and manage the nation's ecosystems, their biodiversity, and the services they provide' (LTER Network, 2011a). Such sites are crucial, because many ecological patterns and processes cannot be fully appreciated or understood by focusing on only a limited spatial or temporal scale. As of 2011, 26 research sites form the LTER network. The first LTER sites were selected and funded in 1980, and the most recent additions have been the three 'coastal' LTERs (Georgia Coastal Ecosystem, Florida Coastal Everglades and Santa Barbara Coastal) set up in 2000. The Baltimore and Central Arizona-Phoenix LTER sites were established in 1997.

The Baltimore LTER site was selected because of a long-standing history of socio-ecological work and good geographical datasets that leant themselves to the interdisciplinary study of urban environments, and particularly the urban region as a system. Baltimore is an urban region with a declining population (broadly around 2.7 million, with the city itself at *c.*620,000, down from a peak of 950,000 in the 1950s). The area has undergone a shift from a manufacturing-based economy to a service-based economy, and has seen a notable amount of urban redevelopment in the last 40 years, along with increased expansion of the urban periphery and abandonment of the urban 'core'. It has a subtropical climate with high precipitation. Research foci at this site include energy and matter fluxes; reciprocal influences of spatial structures (both physical and social) and ecological processes; public understanding of and engagement with urban ecology; and the drivers and impacts of non-native species (LTER Network, 2011b).

The Central Arizona-Phoenix site was selected because the arid southwest of the USA has been urbanising at a very fast rate since the Second World War, and consequently represents a useful site to investigate the effects of urbanisation on society and of society on urbanisation. Phoenix is the capital city of Arizona and the broad urban area maintains a population of over 4 million. The terrain is largely flat and the urban region is generally constructed in an extensive and well-planned grid system, so that the population density is for the most part relatively low compared to many other urban regions. The region receives little precipitation, and access to water resources with growing urbanisation is a key issue for both urban ecosystems and society. Research foci at this site include ecological, social and health relationships with urban climate change; impacts of urbanisation on urban hydrology, water resources and associated services; biogeochemical patterns and processes and their relation to ecosystem services; and urban biodiversity and conservation, and how these are influenced by human activities and behaviours (LTER Network, 2011c).

Both urban LTER sites are multidisciplinary, in that they focus very explicitly on the linkages and feedbacks between ecological or environmental processes and social drivers. Thus far, these sites have been instrumental in demonstrating the relatively slow rate of change between urban greenery and socio-economic status of residential areas, elevated atmospheric CO_2 and air temperature in urban areas, the ways in which watersheds may be used as integrative tools in urban areas, the impacts of urbanisation and the relationship between landscape metrics on plant species richness and ecological quality, and developing the human ecosystem framework (Figure 1.3 on page 15), amongst other things (e.g. Pickett and Cadenasso, 2006; Stiles and Scheiner, 2010). Much more light will be shed on ecological patterns and processes via the use of these urban LTER sites in the coming years.

4   *Future growth.* Much of the world's future population growth will be centred in urban areas (UNFPA, 2007), and urban ecosystems are likely to grow both in spatial extent and intensity in the immediate future. Given the potential for further environmental degradation that this brings, it is clear that achieving a good ecological understanding of urban environments is paramount. In particular this will feed into the development of sustainable urban areas, both via the modification of existing urban ecosystems or the future construction of new urban environments (discussed further in Chapters 7, 8 and 9).

5   *Resource abundance.* Urban regions represent concentrations of resources that can be brought to bear on some of the important ecological questions of our time, including biodiversity conservation, the maintenance of ecosystem services and sustainable development. Natural resources (e.g. biodiversity) can be surprisingly high in urban regions, both in remnant and introduced ecosystems (e.g. Galluzzi *et al.*, 2010; Lugo, 2010; Francis, 2011), and can facilitate the study of ecological and socio-ecological patterns and processes (Pickett *et al.*, 2008). Political and economic power is also often concentrated in urban regions, providing the required financial and political strength to potentially achieve sustained research and action, particularly when the citizenry also benefit. One example of this is the

Figure 1.4 The Cheonggycheon River, Seoul, Republic of Korea. A river previously buried under a motorway has been exposed and converted into a riverside park in the middle of a major urban complex to improve quality of life for residents. (Photograph by In Mook Choi, used with permission.)

rehabilitation of a 6 km section of the Cheonggyecheon River in Seoul, Korea, which had been buried under a motorway in the mid twentieth century, but was exposed and re-engineered to be more aesthetically pleasing and ecologically sound in the early twenty-first century, at an estimated cost of *c.* US$380 million (Figure 1.4; Kim, 2005). Although this rehabilitation was to a large extent performed for political, economic and cultural reasons rather than environmental (Kim, 2005), it nevertheless demonstrates that considerable ecological benefits can result in urban areas with sufficient resource investment. Also, the abundance of human resources (not just people-power but also knowledge) in urban areas should not be underestimated. Motivated citizens, whether amateurs or experts, can contribute substantially to data collection (for example via citizen science, see Francis and Lorimer, 2011; Section 7.6), to conservation enactment, and to decision making at various levels (e.g. Francis and Goodman, 2010).

6   *Legacies*. Urbanisation may perhaps represent one of the few enduring legacies of human civilisations. Historical urban areas form the basis for much archaeology and anthropology because they tell us a great deal about past societies (Gates, 2003). Modern industrial society may leave a lasting legacy of environmental degradation and technological advancement, but the construction of extensive urban areas is surely one of the great achievements of our time, regardless of its negative and positive aspects, and something that is likely to be preserved for future civilisations (e.g. Zalasiewicz, 2008). The transition through to sustainable

and ecologically literate design is likely to be one of the great accomplishments of post-industrial society. Again, such a transition depends upon a sound ecological understanding of these fascinating ecosystems.

1.8 Ecology *in* urban regions and ecology *of* urban regions

There is a useful distinction to be drawn between the study of ecological patterns and processes of ecosystems *within* an urban region, wherein the urban region is essentially the environmental context for the unit of investigation, and the study of urban regions as broad-scale ecosystems in their own right. Most studies in urban ecology focus on the former. The well-known studies of behavioural changes in animals in urban environments, such as the decreased territoriality and communal living of urban foxes (*Vulpes vulpes*), are good examples (e.g. Contesse *et al.*, 2004). These adaptive changes are caused by the urban environment within which the species live, are processes operating within the urban ecosystem, and may tell us useful things about the urban ecosystem; but they are not a study of the system itself. Likewise, many studies focus on ecosystems found in urban environments (such as parks, or walls; e.g. LaPaix and Freedman, 2010, Francis, 2011; see Chapters 4 and 5), but do not consider the more broad urban region ecosystem as a whole. This is in part a legacy of the piecemeal development of ecological studies in urban regions, and also a reflection of the heterogeneity and complexity of such systems (Wu *et al.*, 2011).

More recently, there have been some studies that do examine urban regions as ecosystems, considering their fundamental characteristics, their functioning, and the ways in which they interact with their immediate and global environment (Rebele, 1994; Grimm *et al.*, 2008). In some ways this may be considered an artefact of the spatial scale under consideration: as Rebele (1994, p. 175) notes, 'the boundaries of [an ecosystem] are laid down by the scientist more or less arbitrarily, depending on the topic under study and on practical consideration'. The more holistic consideration of an urban region as an ecosystem in and of itself is important, however – partly because it is only by considering urban areas in such a way that some key questions regarding their ecology may be addressed – and there are increasing efforts to achieve this amongst the urban ecology community, such as the development of the 'urban metabolism' concept (Grimm *et al.*, 2008; Pickett *et al.*, 2011; Section 3.2.2).

Much of the discussion in this book will be of ecological patterns and processes *within* urban regions, as this is (1) the level at which most research and understanding has been developed, and (2) the most useful level for urban conservation and planning that will require an ecological understanding. Nevertheless, consideration of the characteristics of the urban region ecosystem itself is of great importance, and consequently this is explored in more detail in Chapter 3.

1.9 Chapter summary

- Urban ecology is the study of ecological patterns and processes within an urban context.
- Urban ecology focuses on both the ecology *in* and the ecology *of* urban regions.
- Definitions of 'urban' are not standard and vary geographically, but perhaps the best definition is 'populated areas provided with basic services (e.g. homesteads,

electricity and water supply, drainage), where more than 1000 people km^{-2} (>10 inhabitants ha^{-1}) live or work, and an important proportion of the land (>50 per cent), in a "city-scale" (i.e. a spatially delineated area such as a city or town), is covered by impervious surfaces (e.g. buildings, streets, roads)'. The terms 'city' and 'town' are used for political designations and have little ecological relevance.

- Urbanisation occurred slowly around the globe until the industrial revolution of the eighteenth century, when very dense urban regions began to be constructed in Western Europe and then many other areas in the twentieth century, often highly spatially planned (as in North America). Most future urbanisation is likely to occur in Asia, Africa and South America.

- Urban regions are interesting ecologically, as: (1) they are novel environments; (2) they present unique forms of environmental impact and change, and present new challenges to biota; (3) they are replicable systems (within limits), and can therefore be used to test hypotheses at different geographical scales; (4) most of the global population live in urban regions, and this proportion is likely to grow substantially in the coming decades; (5) they are areas where many different types of resource are concentrated; and (6) they are an important legacy for human civilisations.

- The LTER sites at Baltimore and Central Arizona-Phoenix in the USA are important urban regions for long-term data collection, and will provide us with a great deal of useful data over the coming years.

1.10 Discussion questions

1   How is 'urban' defined in your country/region? How useful is this? Would you be able to designate an urban region for management purposes easily, based on this definition? Is it comparable to definitions for other countries/regions?

2   What are some of the implications of increasing urbanisation in Asia, Africa and South America? How is this likely to differ from previous forms of urbanisation?

3   Are you more or less likely to support urban conservation measures compared to non-urban ones, and why?

4   Is the more 'cosmopolitan' biodiversity found within urban regions a good thing or a bad thing? What might be the more positive and negative aspects of this 'recombinant' ecology?

5   Where should other urban LTER sites be located around the world, ideally? Why? What criteria would you use to select candidate sites?

Chapter 2

Urban form, structure and dynamics

2.1 Introduction

Despite being relatively well-planned and therefore spatially organised from a human perspective, urban environments are ecologically complex and heterogeneous in structure. This complexity depends upon the spatial scale considered, so that at the broadest scales particular typologies or patterns of urban form are observable, with implications for the ecology of the urban region. At finer spatial scales, patterns or trends in the arrangement of biotic and abiotic components can be observed (such as the density of buildings, gardens or street trees), again with related influences on urban ecology. Furthermore, urban environments are by nature dynamic, with relatively frequent changes in their physical fabric (e.g. construction and organisation of the built environment) and social organisation (e.g. as groups broadly defined by socio-economic or cultural characteristics shift location over time). Consequently, this chapter focuses on the different forms and structures of urban areas at different spatial scales, from the broad spatial layout and organisation of entire regions (including the formation of urban typologies), to the spatial organisation of landscape components within the urban complex. The biophysical dynamics of urban environments at each scale are also briefly discussed. Although this topic is vast and could form the basis for an entire book, and as such one chapter cannot provide comprehensive coverage, a broad overview of spatial organisation in urban regions is essential for developing an understanding of urban ecosystems and their ecology.

2.2 Defining urban form and structure

The terms 'urban form', 'urban structure', 'urban morphology' and 'urban fabric' are often used interchangeably, and can refer to the spatial organisation of both physical and social parameters, such as housing density, arrangement of transportation networks, patterns of land use, population density or employment distributions (see, e.g. Tsai, 2005; Bramley *et al.*, 2009; Tian *et al.*, 2010). This chapter focuses on the biophysical components, rather than socio-economic or cultural characteristics, of urban regions, though of course the two are closely linked and some cross-discussion is required. 'Urban form' is considered here to refer to the broad spatial characteristics of an entire urban region. This includes the shape and arrangement of the urban complex(es) that represent urban land use within the region. 'Urban structure' is the spatial organisation of the urban complex at finer spatial scales (landscape to individual

Box 2.1 Patch–corridor–matrix model

The discipline of landscape ecology focuses on interactions between landscape form and process, and how they may relate to landscape functioning. It has been dominated by the patch–corridor–matrix model of landscape structure since the model's initial publication in Forman and Godron's (1986) seminal text *Landscape Ecology* and its further substantive development in Forman's (1996) *Land Mosaics*. The model is essentially an anthropogenic simplification of landscape structure (regardless of scale) into a mosaic consisting of the background land use (termed 'the matrix'), which is often the most abundant form of land use for an arbitrarily delineated landscape, and into which are embedded discrete areas of other land-use types, which are termed 'patches' if they are broadly non-linear and 'corridors' if they are linear (relatively long and thin). 'Patch' and 'corridor' are essentially descriptive terms for spatially defined ecosystems (e.g. a patch of woodland, a river corridor). They are also landscape components, i.e. spatially discrete areas that together make up a landscape. Thus, 'patch', 'ecosystem' and 'landscape component' are terms often used interchangeably within landscape ecology.

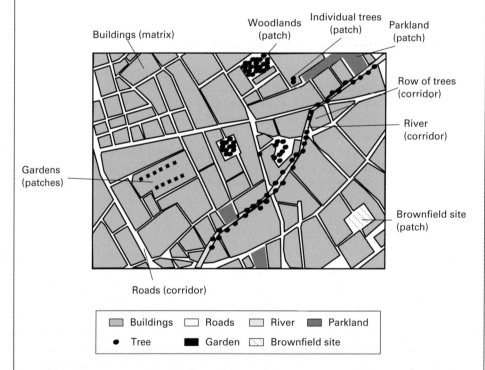

Figure 2.1 An example of the application of the patch–corridor–matrix concept to an urban landscape. Buildings represent the greater land cover or 'matrix', with patches and corridors of other land surface type embedded within it. This way of conceiving the urban landscape has become a cornerstone of ecological investigations and urban planning.

Within an urban context, the built environment may generally be considered the matrix, while lakes, ponds, gardens, woodlands, brownfield sites and so on would be patches, and rivers, canals, roads and railway lines (for example) would be corridors (Figure 2.1). The patch–corridor–matrix model is very useful, and has been the basis for much landscape investigation in urban areas, as well as the application of landscape metrics, and can be applied at any spatial scale – although usually landscape ecology is applied at the scale of kilometres, the principles can be applied at any scale. Patches and corridors do not both have to be present, though a matrix does – but this is often neglected and landscapes are described simply as mosaics of different patch types. Although the patch–corridor–matrix model is not always ideal, it is very useful for illustrating urban spatiotemporal patterns, which is an important first step in developing an understanding of urban ecosystems.

streets and buildings) and is essentially how various urban components are arranged (see Box 2.1). The individual components are streets, gardens, parks, rivers, canals, buildings and the other many and varied types of ecosystems that are found within urban areas; these components are discussed in more detail in Chapters 4 and 5.

2.3 Spatial organisation of the urban region

At the very broadest (regional) scale, particular forms or typologies of urban spatial organisation can be observed (e.g. Luck and Wu, 2002; Forman, 2008). The two-dimensional plan form or 'shape' of an urban area will reflect patterns of past and present urbanisation, and will depend upon the geomorphology, ecology and climate of the region, as well as historical and current trends in urban planning (Luck and Wu, 2002; Dietzel et al., 2005; Forman, 2008). Opportunities and constraints created by the physical environment can have an over-arching influence on urban form at the regional scale, and will have implications for ecological patterns and processes.

2.3.1 Geographical features and spatial organisation

The most dramatic influences result from the presence of major water features (large rivers, lakes), coasts and mountains (e.g. Kasanko et al., 2006; Papanastasiou and Melas, 2009). Because such features represent useful resources and are often prime reasons for the original establishment of urban settlements, and because they may preclude urban development to a large extent, they help to shape the urban form and in some cases impose limitations on lateral expansion and/or densification of the urban complex. For example, part of the reason for the compact, high-density form of Hong Kong (with an average density of around 26,600 people km^{-2} and a maximum of 116,500 people km^{-2} in the year 2000), is the situation of the region between the coast and the more mountainous hinterland that precludes lateral expansion for the most part, and led instead to a historical focus on densification of the urban complex (Jim, 2002).

In situations where a large river runs through the urban region, this essentially bifurcates the region and may represent both a substantial barrier to, and conduit for, the movement of both biota and abiota. Often the major river in the region is located relatively central to the urban complex, as urbanisation has occurred mainly around the river, with some of the densest development adjacent to it. Some of the most notable examples of this situation are London (UK), Paris (France), Seoul (South Korea), Moscow (Russia) and Manila (Philippines). Such large urban rivers may form barriers or filters to species movement, particularly because they are often heavily channelised with vertical or near-vertical flood embankments, and lack riparian habitat or any easy crossing points. Combined with the high-density built environment that surrounds them, this can effectively prevent movement of some species within or through the urban complex, even at relatively narrow widths (<50 m) (Tremblay and St. Clair, 2009, 2011). As conduits, the water flowing through the region can be an important resource, whether for direct use (drinking, irrigation, industry) or indirect use (fishing, recreation, waste removal). However, rivers also conduct pollutants both into the urban complex (e.g. from agricultural areas or even other urban regions further upstream – a problem further exacerbated by the tendency of urban regions to be located in the middle or lower sections of river catchments) and out of it, often with an increased pollutant load (e.g. Taylor and Owens, 2009; Mohiuddin *et al.*, 2010). They may also represent a significant flood risk, often leading to extensive hard engineering and flow management (Francis, 2009a). Larger rivers are frequently fed within the urban complex by smaller tributaries, though these do not feature as heavily in patterns of development (though see below and Chapter 4) and are often simply incorporated into sewerage infrastructure (Elmore and Kaushal, 2008). Some urban complexes may, however, be located at the site of a major tributary joining the main river channel, and in such situations further bifurcation of the complex may occur. Some good examples of this situation include Taipei (Taiwan), Montreal (Canada) and Lyon (France).

Some urban regions are based around lakes (well-known examples being found in Switzerland and the US/Canada Great Lakes region), and in such cases the urban complex may expand along the lakefront and take a broadly crescentic form, or establish all around the lake and essentially enclose it, depending upon the size of both lake and urban complex. In these cases, the lake acts as a focus for the region, and may be a substantial resource for transportation, trade and recreation; development and industry will typically be densest, at least historically, adjacent to the lake front. Such lake-adjacent development leads to the destruction of riparian habitat, and urban lakes may also be sinks for pollutants with limited capacity for fast natural recovery (see Chapter 4).

Many urban regions are located in coastal areas due to the resources, trade and defensive opportunities that such locations offer, and many of the largest global urban regions are coastal (Figure 1.2; see also Martinez *et al.*, 2007). Coastal forms expand both inland and along the coast, often forming broadly linear, semi-circular or crescentic patterns with the coast as the area of historically densest development. Such regions may be potentially susceptible to flooding or coastal erosion, and so flood defences are often constructed, sometimes extending along estuaries. Urban coastal ecosystems are often degraded, in part due to frequent and intense resource harvesting, coastal traffic, and pollution (e.g. Feng *et al.*, 1998; Cetin, 2009). Coastal regions are particularly at risk from any sea-level rise (e.g. due to climate change) and may also be vulnerable to altered urban hydrological cycles or abstraction that depletes groundwater, as this may cause geologically young coastal strata to subside (Nicholls, 1995).

Mountains also present significant barriers to urban development and consequently shape urban regions. Although many mountain settlements exist (mountains can offer some key resources, including defensibility), they rarely grow into sizeable urban areas due to an absence of stable land to build on, and mountains often 'contain' urban regions (e.g. Hong Kong). Mountainous or hilly urban areas are also potentially more vulnerable to natural hazards such as slope failure and landslides (Alexander, 1989), further discouraging both extensive and intensive development in many cases.

What is clear is that features of the earth's surface can have an effect on patterns of urbanisation and consequently urban form, and combinations of barriers such as rivers and mountains may further complicate both form and structure. These examples remain only broad generalisations, however, and it is important to remember that every urban region is somewhat unique. Nevertheless, discussion of such general forms is useful when considering their ecological implications, and as a basis for further investigations.

2.3.2 Simple urban forms

In situations where geographical barriers and boundaries are less prevalent, the urban region may take various forms as described by Forman (2008; see Figure 2.2), though these are necessarily simplifications of the real world and many more specific and varied forms exist for particular regions. These forms may be classified into those created by urban expansion that is spatially *adjacent* to the urban complex, and that which is *removed* from the complex. Perhaps the simplest basic form of adjacent expansion is the 'concentric rings' model, in which reasonably dense expansion often occurs more-or-less concentrically at the boundaries of the urban complex, though often not consistently, so that a somewhat irregular pattern may emerge. This is often due to development taking place on segments of land extending outwards into the countryside from the edge of the complex, in stages. Such segmented development may be due to (for example) the construction of major transport links, immigration by particular ethnic groups who establish in locations with a strong cultural identity (including informal 'slum' settlements), or governmental policies that focus or guide development into certain areas or land-use types (e.g. Masek *et al.*, 2000). Conversely, growth in particular areas may be prevented by the presence of protected land, usually green space, which may lead to the formation of 'wedges' (following Forman, 2008) of green space in the urban complex, or a green belt surrounding the complex (e.g. Amati and Yokohari, 2006). Historically, many urban areas would have maintained relatively simple forms of more-or-less concentric expansion, still seen in some major urban regions (e.g. Tian *et al.*, 2010), though this model is now considered too simplistic in many cases (Warren *et al.*, 2010; Pickett *et al.*, 2011). Certainly it should not be assumed that such expansion creates circular urban complexes, as a wide range of shapes are found (e.g. Bento *et al.*, 2005).

A similar, though more complex, form of adjacent expansion is the dispersed-sites model (Figure 2.2), wherein more diffuse, spontaneous and often low-density

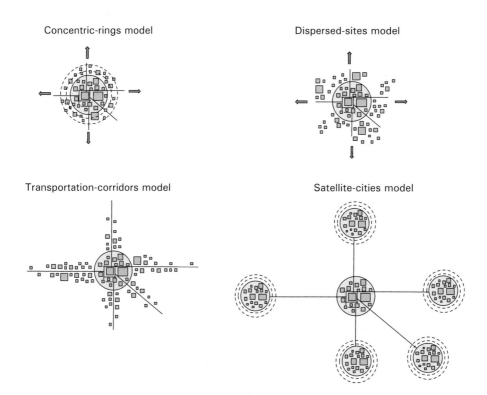

Figure 2.2 Four simple urban region typologies observed by Forman (2008). The 'concentric-rings' and 'dispersed-sites' models may be considered 'adjacent expansion', while the 'transportation-corridors' and 'satellite-cities' models may be considered 'removed expansion'. See text for further details. (Modified from Forman, 2008; reproduced with permission.)

development occurs in patches in exurban areas and thereby becomes assimilated into the urban complex, connected by often extensive and possibly high-density infrastructure networks (and is perhaps most characteristic of the derogatory term 'sprawl'). This creation of less-dense urban patches in the exurban area is nevertheless considered amongst the most ecologically damaging of processes, as it reduces the area and connectivity of the natural or semi-natural ecosystems existing prior to urbanisation (Radeloff *et al.*, 2005; Forman, 2008), essentially leading to habitat fragmentation (Wu *et al.*, 2011).

As with the concentric rings model, the specific location and extent of such expansion will depend on the various socio-economic and geographical factors driving urbanisation. Kasanko *et al.* (2006) have observed that although many European urban complexes have experienced dispersed expansion since the 1950s, the extent varies geographically with Southern European complexes being more compact and those in Northern and Eastern Europe being more dispersed, though the trends are converging. In China, rapid expansion is creating a blend of urban and non-urban land use, termed *desakota* (*desa* = village and *kota* = town, in Bahasa Indonesian), which is relatively unique to some regions (McGee, 1989; Xie *et al.*, 2006).

Both of these adjacent expansion forms generally exist where an urban region has developed around a central nucleus. For the last two centuries the majority of urban regions developed structures that were based around a spatially explicit economic 'hub' that often combined relatively high building densities with high population densities and formed the location for key financial activities (sometimes referred to as a Central Business District or CBD; see Box 2.2). Some particularly well-known examples include the City of London (UK), Manhattan in New York (USA), and the Central District of Hong Kong (China). As noted above, the urban environment tended to become less dense with increasing distance from the historical CBD as a general rule (Anas *et al.*, 1998). This classical 'monocentric' pattern of urban form has become less common in the last century. Increasing urbanisation and the development of increasingly effective infrastructure has led to the development of 'polycentric' spatial patterns in many urban regions, with economic hubs established in different locations, some of which may have been pre-existing smaller urban areas that were

---

Box 2.2 Economic drivers of urban spatial form and structure

Much work on the spatial structure of urban regions has focused on the distribution of population density and economic variables, such as employment, income, or type of industry. Although the focus of this text is on ecology, the roots of much urban quantification in economic geography are worth some further scrutiny, as these variables do have some ecological relevance – denser populations for example are usually associated with a denser built environment, which in turn often leads to reduced biodiversity. Although these links are not as strongly researched as we might wish, some general consideration is useful.

The drivers of urban formation and growth (and decline) are essentially economic. The initial establishment and growth of many settlements was at geographical locations where resource harvesting, manufacture or transportation was made possible or most effective, which is why so many urban regions are centred on waterways and coasts, at least for those urban regions that were settlements prior to the development of mass transportation in the nineteenth century. Prior to the development of mass transport (railways, motor cars and major roads), the difficulty of transportation within a region (i.e. not via the coast or waterways to other regions), which was mainly achieved via horse and cart, led to the tendency for populations to aggregate around a focal point for the urban industry (i.e. around the port location, or a manufacturing area). Residences, industry and businesses would all be found close by each other – although some segregation may occur (with jewellers located along a single street or district, for example), the urban complex was spatially limited, densely populated and compact. This is what would now be known as a Central Business District (CBD); only the poorer people would live away from the district, having farther to travel and less convenience. In some of the ancient and well-established urban centres of Europe, these trends can still be observed, with CBDs containing a mix of residences, businesses and workshops.

In more modern urban regions (such as many cases in the USA) and following the development of mass transport and communication technology, settlements formed at key points along transportation networks, such as major railway or motorway intersections or terminals. Urban regions have grown around shipping and trade based at such locations, though again usually with a single focal point (CBD). As more individual and intra-urban transportation become possible (with the development of electric trams and eventually the motorcar, rail and underground rail systems), affluent people moved away from the (often unpleasant) CBD to live in the less densely populated edge or exurban areas. The CBD therefore often became less residential and densely populated in many areas, especially fast growing or relatively newly constructed urban complexes, though they also expanded in size as the urban complexes grew. In the early part of the twentieth century, the CBD manufacturing bases still generally relied on the waterway, coastal or rail links, and the CBD remained in the same location. By the mid twentieth century, transportation (particularly road and rail) networks had developed to the extent that manufacturing industry was able to relocate to the cheaper and more easy to access exurban or urban edge areas and the CBDs became generally service based, as is the case with most large urban complexes today.

This model generally follows the principle of monocentricity, with one focal point (the CBD), but the changing possibilities of transport and communication systems have developed 'polycentric' situations where several centres may emerge, or urban complexes may expand to join together, with several CBDs (as in many urban conurbations). In some cases, the CBD develops in a linear fashion with urbanisation, creating a Central Business Area; this has been observed particularly in many Latin American regions (e.g. Vandersmissen et al., 2003). Modern, extensive, urban regions are particularly complex, but the past and present socio-economic structure has relevance to the physical and social fabric that urban regions display.

amalgamated into the larger region as it grew. Increasing globalisation of economic activities, tied less to specific spaces following the information revolution, has led some urban areas to move beyond polycentricity, with economic centres becoming less explicit in recent decades (Anas et al., 1998).

This polycentric trend has also contributed to the development of urban conurbations, or urban regions that have combined or overlap spatially, which relates to the first form of 'removed expansion'. This is the 'satellite' formation, where nearby urban regions (which can themselves be of a substantial size) become part of the larger conurbation via the development of major infrastructure routes, such as motorways or train lines (Figure 2.2). These allow people to commute from the satellite areas to the primary urban complex and may help to preserve natural or semi-natural exurban areas by focusing development elsewhere, though the creation of major transport infrastructure inevitably has some level of ecological impact on the exurban landscape (Forman et al., 2003; Roedenbeck et al., 2007; Glista et al., 2009). The second form of removed expansion is the transportation-corridors model, in which urban development occurs along the major transport routes into an urban region, whether these connect to satellite regions or not. In this form, development is therefore not focused primarily in the satellite areas but rather along the connecting infrastructure (Figure 2.2). This will also have ecological impacts, including increasing degradation of habitats adjacent to the road and an increase in detrimental edge effects (see Box 2.3). For example, negative impacts from traffic pollution, noise, species invasions and so on can affect vegetation health, animal movement and species assemblage composition up to hundreds of meters from the road edge (e.g. Forman and Deblinger, 2000; Bignal *et al.*, 2007).

The essential ecological difference between adjacent and removed expansion is the type of impacts they may have. Adjacent concentric expansion is less likely to degrade and fragment the exurban landscape in general (Tratalos *et al.*, 2007a), but will have localised effects within the area of expansion and may exacerbate densification within the complex. Adjacent dispersed-sites expansion may also limit the spatial extent of impacts, but will increase fragmentation over a larger area than concentric expansion. Removed 'satellite' expansion may create impacts away from the primary urban complex and degrade the region as a whole. Removed transportation-corridors expansion will fragment the habitat next to and between infrastructure, and may increase detrimental edge effects in these areas. Such fragmentation has been shown to have a substantial negative impact on ecosystem structure, processes and diversity, and is a

---

Box 2.3 Edge effects

'Edge effects' relate to a range of characteristics associated with the edges of landscape patches and corridors, and essentially highlight how the edge environment differs from the more interior environment of the patch/corridor. In essence, the surrounding matrix will have extend its influence for some distance into the patch/corridor, and vice versa. So, for example, a woodland patch surrounded by an open field will experience greater light and wind penetration in the edge area compared to the interior, while exerting a shading effect on the field. The woodland edge would also experience an increased abundance of deposited seeds and sediments due to the drop in wind speed, a trend termed impaction. These edge environments therefore present different resource abundances and niches to the interior of the patch/corridor or the matrix and may consequently support different species assemblages. These edge environments, where conditions change rapidly over a relatively short distance, are often termed 'ecotones' (Hufkens *et al.*, 2009). In highly fragmented urban landscapes, patches may be small and corridors thin, and as a result have a high proportion of edge. This may contribute to the abundance of edge-tolerant species and a reduction of interior species in urban areas (see Chapter 6). Edges have also been linked to the establishment of non-native species (e.g. Hunter and Mattice, 2002; Hansen and Clevenger, 2005). Note that establishing the extent of edge effect varies depending on the process or characteristics being considered – quantification of the woodland 'edge' as affected by wind penetration may be different for that related to foraging use by small mammals, for example.

major ecological concern (Laurance, 2008). However, specific impacts will depend on the pre-urban land use, and the relative merits of adjacent or removed expansion are uncertain, and will vary for different species (e.g. Gagne and Fahrig, 2010; Fontana *et al.*, 2011).

Of course, these broad forms are generalisations and are not necessarily mutually exclusive; a given urban region may display one or more aspects of any of them at any given point in time, though generally one dominant pattern will emerge. Particularly large and complex conurbations, which incorporate many interlinked urban regions, may result from extensive adjacent and removed expansion and display none of these obvious forms due to their large spatial coverage and complexity. In their analysis of the urban expansion of Washington DC from 1973 to 1996, Masek *et al.* (2000) note that adjacent expansion around the Washington metropolitan area was accompanied by remote expansion in small towns and transportation hubs, thereby displaying evidence of more than one form. Likewise, Luck and Wu (2002) noted that a distinct 'form' for the much-studied Phoenix (Arizona, USA) region was also unclear from their analyses of landscape patterns. Nevertheless, consideration of such forms may be illuminating when examining ecosystem processes and environmental impacts, and comparing between regions.

2.3.3 Spatial characteristics of urban regions

Although all urban areas are unique to some extent, they generally display some common spatial characteristics at the regional scale. The most extensive review of characteristics comes from Forman (2008) and his surveys of 38 major global urban regions, but many other studies support these general observations for individual regions:

1 *Urban regions are a mix of green space and the built environment.* Most urban regions contain natural or semi-natural areas near or within the urban complex, generally termed 'green space', which may essentially be defined as vegetated undeveloped (or less-developed) land, though the lack of development is somewhat relative – agricultural areas for example may be included but are not considered 'undeveloped' from an ecological point of view (see Chapter 4 for further discussion of green space). This highlights the heterogeneity of urban regions and an abundance of ecosystem types. Consequently the built environment, although dominant, does not form comprehensive cover, and green space can cover extensive areas (Fuller and Gaston, 2009). Most green space is in the form of parks and gardens, with woodland generally representing the largest patches of green space, followed by parkland, brownfield sites and gardens. Patches of green space generally become smaller and denser within increasing coverage of the built environment, which often peaks in or around urban centres. Numerous studies have been performed into the ecological value of urban green spaces and green corridors, green networks or green belts, demonstrating positive benefits relating to biodiversity and species population persistence, carbon sequestration, pollution mitigation, reduction of the urban heat island effects, habitat provision, and improved aesthetics and amenities (e.g. Bengston and Youn, 2006; James *et al.*, 2009; Dallimer *et al.*, 2012). This is explored in more detail in the following chapters.

2   *Urban regions support major transport infrastructure.* They usually maintain either major orbital roads (often motorways) around the urban complex, or radial roads extending from the complex like spokes from a hub. Forman (2008) found that the distribution of styles was roughly 50:50, though orbital roads were more common in Europe. Orbital roads have been demonstrated to have notable ecological impacts, including in particular the relative separation of urban/exurban species populations and interruption of gene flow (e.g. Gerlach and Musolf, 2000). Radial roads tend to facilitate road-adjacent development in a strip formation, as described for the transportation-corridors model (Figure 2.2), which may also fragment habitat and interrupt flows of species and materials (Forman *et al.*, 2003; Jackson and Fahrig, 2011). Many regions also maintain railway infrastructure that may have similar impacts, though are generally less severe than for roads (e.g. Tremblay and St. Clair, 2009).

3   *Freshwater ecosystems in urban regions tend to be particularly degraded.* Although rivers are often not 'destroyed', they may be covered by roads or pavements, converted into sewerage or drainage channels, or at the very least heavily engineered and regulated, with accompanying loss of hydrological dynamism and ecological quality (Petts *et al.*, 2002; Gurnell *et al.*, 2007; Elmore and Kaushal, 2008). Forman's (2008) surveys found that most urban rivers have less than 33 per cent of riparian vegetation cover, while urban lakes either have very high (90–100 per cent) or very low (<30 per cent) riparian vegetation coverage. Large rivers, which are often central to the urban complex and are one of the reasons for the initial establishment of a settlement in that location, tend to be amongst the most engineered and degraded of urban rivers. Wetlands (in particular marsh or swamp ecosystems) are among the first natural ecosystems to be lost when urban development occurs, and are often filled in to allow construction (or to prevent disease, e.g. mosquito-borne malaria), or used as landfill sites, particularly when such wetlands have historically been viewed as unproductive (e.g. Pauchard *et al.*, 2006). Forman (2008) further notes that wetland size and presence is linked to population size, with urban complexes of greater than 8 million people retaining no major wetlands.

4   *Agricultural land is often abundant in exurban areas.* Outside of the urban complex, much of the land in many urban regions is agricultural (e.g. Kasanko *et al.*, 2006), in particular arable land, reflecting the origins of settlements where productive soils were found. The urban region therefore often incorporates agricultural ecosystems, though the focus of this book is most specifically on the urban complex, and so these systems are not discussed in detail here.

5   *Topography is often linked to green space.* For those urban regions that have hilly or mountainous terrain within or near to the urban complex, slopes that face the complex often have high vegetation cover, reflecting both the benefits that such vegetation provides (in terms of soil stabilisation and retention, water interception and storage, recreation and aesthetics, for example) and the difficulties associated with building on slopes. For some urban regions that maintain few areas of high elevation, such slopes can be more heavily built upon or degraded, presumably because such locations are relatively uncommon and may provide benefits such as a better local environment and aesthetics, such as fine views.

6   *Gradients of urbanisation.* Urban regions usually display increasing gradients of urbanisation from the exurban area to the urban centre, with associated trends in

landscape patterns, such as patch size, density and landscape complexity. This is discussed further in Section 2.5.

7   *Scales of patterning.* There are scales of patterning within the urban region, with for example ratios of transport network size and connectivity that are suggestive of self-organisation. This is discussed further in Chapter 3.

The above discussion has provided an overview of some of the key factors that may shape urban regions. Urban forms and characteristics are of course closely linked to regional dynamics. These are now considered in more detail.

2.4 Dynamics of the urban region

The dynamics of broad-scale urban development have often been conceptualised as 'waves' of change occurring in phases or cycles, rather than linearly in space and/or time (e.g. Dietzel *et al.*, 2005; Ulfarsson and Carruthers, 2006). Although it is difficult to fully encompass the complexity of the urban region in a single model, a useful conceptualisation of urban spatio-temporal dynamics is the diffusion–coalescence cycle described by Dietzel *et al.* (2005). The cycle occurs in two broad phases, which are 'diffusion' and 'coalescence' (see Figure 2.3; Dietzel *et al.*, 2005; Xu *et al.*, 2007). Diffusion is the expansion of the built environment and urban land use, either immediately adjacent to the urban complex or within the exurban area (and variously termed, for example, 'edge expansion' and 'spontaneous growth' respectively (e.g. Xu *et al.*, 2007); these also relate to Forman's (2008) concentric rings and dispersed sites models; Figure 2.2). Coalescence is the 'infilling' of less developed or undeveloped gaps within the urban complex (Figure 2.3). All three developmental processes (edge growth, dispersed growth and infilling) occur concurrently, but the abundance varies within the urbanisation cycle.

Dietzel *et al.* (2005) hypothesised that the two phases must peak at different times, with the diffusion phase occurring first within a given area, followed by a coalescence phase that effectively 'saturates' the urban complex with built components, though these of course may include embedded green spaces. Further urban growth then stimulates a further phase of diffusion, and so on. The stage of a given urban location within the cycle was also hypothesised to depend to some extent on distance from the centre of the urban complex, with individual cycles generally taking longer to complete with increasing distance from the complex centre (i.e. the wave of urbanisation slows as it expands from the centre). Dietzel *et al.*'s (2005) work in California (USA) provided support for the hypothesis and several other studies have supplied empirical evidence for the diffusion–coalescence cycle, though inevitably its generalities have been exposed. For example, although phases have been observed, they are not always as clear as the hypothesis might suggest. Xu *et al.* (2007) found that for Nanjing (China) urbanisation between 1979 and 2003 occurred in three steps, with edge-adjacent expansion being dominant between 1979 and 1988 (59 per cent of growth area) followed by dispersed growth (27 per cent) and infilling (15 per cent). From 1988 to 2000, infilling increased to 45 per cent, edge expansion to 45 per cent and dispersed growth to 10 per cent. From 2000 to 2003, edge expansion increased to 75 per cent, dispersed growth dropped to 5 per cent and infilling also declined to 20 per cent. This essentially represents a diffusion-coalescence cycle – although expansion occurs

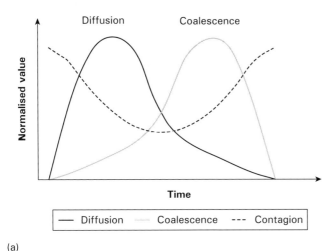

(a)

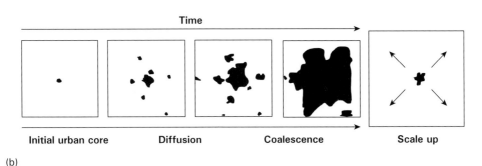

(b)

Figure 2.3 (a) The diffusion–coalescence cycle of urbanisation, whereby expansion and densification of the built environment occur simultaneously, but have a sequential dominance in distinct phases. See text for further details. 'Contagion' is a measure of how large and contiguous patch types are: a landscape composed of few large patches will have a high level of contagion, while a fragmented landscape will have a low level of contagion. (b) The spatial pattern of urbanisation resulting from the cycle. (Adapted from Dietzel *et al.*, 2005; © Taylor & Francis.)

consistently, the coalescence phase is characterised by a relative increase in infilling activity, as outlying urban areas are consolidated and green patches incorporated into more urban land use. Consequently, as urbanisation occurs, less dense urban areas within the urban region may be developed and made denser, so that as the region grows or ages a densification of the urban form takes place. Similar observations on these processes have been made for other regions (e.g. Xie *et al.*, 2006; York *et al.*, 2011; Shi *et al.*, 2012).

The increasing duration of the cycle with distance from the urban centre may seem at odds with the rapid expansion of urban areas documented. Expansion is usually couched in units of distance or area over time, such as the metropolitan area of Washington DC growing by an average of 22 km^2 per year during 1973–1996, accelerating to an average of 40 km^2 per year during 1985–1990 (Masek *et al.*, 2000). Forman (2008), however, notes that a more appropriate measure of urbanisation is the relative growth rate of the urban complex or urban region; that is, the speed with which an initial area increases by, for example, 50 per cent, 100 per cent or 200 per cent. It is important to remember that such growth does not occur in a linear fashion but in phases as noted above, and to avoid judging short-term dramatic rates of increase at face value. A focus on areal expansion alone also does not always account for the ecologically important processes of infilling. Rapid urbanisation tends to favour diffusion over coalescence, tends to be more poorly planned, and also allows less time for the adjustment of ecological processes and communities to environmental changes, particularly resulting from ecosystem fragmentation (Forman 2008; Ramalho and Hobbs, 2012).

Schneider and Woodcock (2008) have further demonstrated four typologies of expansion, including:

1   Expansive growth, wherein areal extent of the complex is large and population density relatively low, with sustained growth on the edges of the complex. This typology was often observed in urban regions of the USA.
2   Frantic growth, wherein expansion is dramatic (6–7 times the median rate of urban regions surveyed) and results in relatively unregulated development. This was observed in urban regions in China, but is not necessarily representative of the country as a whole.
3   High growth, where there is rapid expansion combined with infilling. Examples are Calgary, Bangalore and Hanoi.
4   Low growth, wherein infilling has taken place without notable expansion. Examples include Belo Horizonte, Madrid, Prague and Nairobi.

These remain very broad generalisations and as noted above, both geographical and socio-economic factors may complicate, contradict, interrupt or obscure cycles of urbanisation, such as areas of land that may not be easily developed due to physical characteristics, or because of development restrictions (e.g. Xu *et al.*, 2007). Indeed, expansion is also strongly linked to planning policy and restrictions (or lack of them). Regional urban planning policy may have focused on a 'compact city' strategy, favouring densification of the urban complex, or a 'dispersed city' strategy, encouraging exurban expansion (e.g. Catalan *et al.*, 2008). For example, much of the expansion in Washington DC from 1973 to 1996 was in Northern Virginia, which had relatively lower taxes and less proscriptive zoning laws, allowing greater development, particularly of commercial centres (Masek *et al.*, 2000). In contrast, Maryland purposefully regulated development during the same period, so that it occurred mainly along existing transportation corridors to prevent sprawl (Masek *et al.*, 2000). Consequently, there will almost always be undeveloped or lesser developed areas within the urban complex that complicate simple models of broad-scale urban dynamics (Dietzel *et al.*, 2005).

What is clear is that urbanisation is dynamic, complex, and accelerating globally and (in many cases) regionally. This acceleration of growth is one of the reasons that urban ecology has become a discipline of interest, as the ecological effects of changing land use become more apparent. At the regional scale, urbanisation has been linked to biodiversity and ecosystem decline, water abstraction and depletion, changes to weather and climate, increased pollutant levels and the spread of invasive species, among other things. These impacts are considered in more detail in Chapter 3. Perhaps the most ecologically meaningful examinations of urban structure and dynamics have been conducted at the landscape scale, i.e. patterns within the urban complex.

2.5 Spatial organisation of the urban landscape

The biophysical landscape of urban regions is often characterised using Forman's (1996) patch–corridor–matrix model (see Box 2.1), so that green spaces are considered distinct patches or corridors within a matrix of constructed artificial surfaces such as buildings and roads (e.g. Breuste et al., 2008; Forman, 2008; Connery, 2009). Urban landscapes are arguably one of the most distinct in terms of drawing spatial contrasts between a relatively hostile matrix (the built environment) and remnant or introduced patches (green space) – it is easy to see where pavement stops and a patch of lawn starts, or to contrast a block of houses with a park.

Within the urban complex, the degree of urbanisation and the density of different patches, both introduced and relictual, built and vegetated, will vary spatially according to three broad spatial scales: (1) at the overall complex scale, there is often, for example, a tendency for the densest built environment to be present in the core of the urban complex, with this reducing with increasing distance from the centre and into the exurban area, with a consequent increase in green space (Peiser, 1989; Luck and Wu, 2002); (2) within the urban complex, type and density of different patch types change according to the use of a given area (e.g. for industry, business, residences; Akbari et al., 2003; Baker and Harris, 2007); (3) within specific zones (e.g. residential) there will be notable differences in patch diversity and abundance, relating to the patch sizes of buildings and associated features such as gardens, as well as the varying impacts of human activities and management (e.g. Arnold and Gibbons, 1996). These are very simplified, general patterns and will vary substantially between different urban complexes (and relate to socio-economic factors as discussed below), but nevertheless help to illustrate the level of spatial complexity present in urban landscapes.

In the last two decades there have been attempts to quantify the spatial patterning and complexity of urban regions, and in particular the urban complex, using landscape metrics (e.g. Schwarz, 2010; Wu et al., 2011; Angel et al., 2012). Each urban region has its own unique aspects relating to its structure and organisation, but some general trends can be surmised from case studies conducted (e.g. Luck and Wu, 2002; Andersson et al., 2009; Wu et al., 2011). Generally patterns relate to the level of urbanisation of a particular area. Broadly, increasing urbanisation (in terms of construction of the built environment, usually with population growth) leads to increasing fragmentation of the landscape, with patches of different land-use types becoming smaller and more densely arranged (Luck and Wu, 2002; Wu et al., 2011). This is reflected in, for example, large patches of woodland, grassland, agricultural lands or wetlands being built upon piece by piece with the concomitant construction of infrastructure such as roads, that divides the original patches into smaller and smaller segments. At the same time, land-use patterns become more evenly distributed, so that distinct transitions between (for example) densely and less densely built areas, woodlands, agricultural land and so on, become blurred, and (with the exception of perhaps the oldest and most developed areas), different land-use types can be found throughout the urban complex. Overall landscape complexity also increases, with the perimeter–area ratio of the landscape mosaic increasing and more complex patch shapes emerging with greater urbanisation (e.g. Wu et al., 2011), though this trend has not been found in all studies (e.g. Andersson et al., 2009), further highlighting the individuality of urban regions.

Often these changes to landscape metrics will emerge exponentially over relatively short periods after some stability, following an economic or social shift that increases population and construction (e.g. during a diffusion phase). Wu et al. (2011) for example demonstrate that metrics such as patch density (as a measure of fragmentation) remained stable from the early part of the twentieth century until the 1950s–60s for Phoenix and Las Vegas, whereupon these regions experienced rapid growth. Although the socio-economic drivers of this growth were different for the two regions (industry development, retirement and tourism for Phoenix and gambling/entertainment for Las Vegas), similar patterns of urbanisation emerged, possibly due to their similar natural environments and stages of urban development (Wu et al., 2011).

Another way of conceiving spatial patterns in the urban landscape is to examine gradients of land-use characteristics and landscape metrics across an urban region. If an imaginary line is drawn across the region, in theory it would be expected that land use would broadly change from the original land use (whether natural, semi-natural or, e.g. agricultural) to less dense urban cover, to dense urban cover in the more central part of the complex, back to less dense cover, and then back to original land use (Figure 2.4). This would usually relate to fewer, less dense, larger patches in the exurban areas, moving to more, smaller, more densely arranged patches towards the urban centre. This gradient may be interrupted by occasional larger patches of green space that may reflect, for example, preserved areas or geographical barriers that have made urbanisation difficult (e.g. Jim, 2010). Luck and Wu (2002) found this broad pattern for Phoenix, Arizona (Figure 2.4), though of course in some respects patterns are not symmetrical (in Phoenix distinct 'rings' of land use or development were absent, for example), reflecting the complex and irregular shapes of urban regions as well as geographical constraints on land use and development. However, Ramalho and Hobbs (2012) caution against the use of such gradient concepts to generalise on urban form and structure, noting that spatial patterns do not always relate to theoretical or perceived gradients and so assumptions regarding patterns and position within the complex relative to centre and edge are not always appropriate.

At the landscape scale, patch and corridor configuration is important in influencing ecological patterns and processes throughout the region. Two key factors that relate to configuration are connectivity and heterogeneity.

2.5.1 Landscape connectivity

Much urban landscape ecology is based around the principle that a more connected landscape (that is, less fragmented and more contiguous) will have more robust and integral flows of organisms and materials, and thereby maintain more natural (for the ecosystem in question) levels of biodiversity, ecosystem processes and ecosystem services (e.g. Fernández-Juricic and Jokimaki, 2001; Tremblay and St. Clair, 2009; Shanahan et al., 2011). The importance of connectivity has been established in several

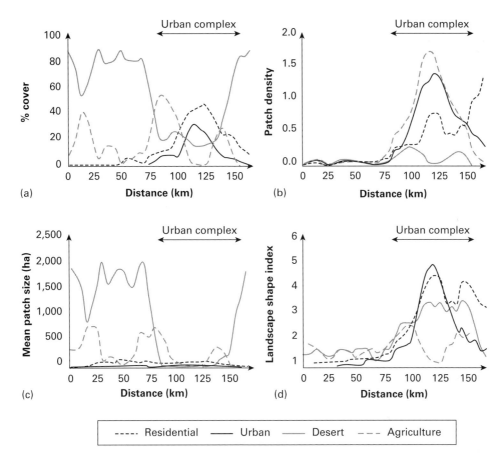

Figure 2.4 Landscape metrics observed along a gradient within the urban region of Phoenix, Arizona. Metrics were based on patch measurements, with patches classified as urban (e.g. retail, industry, business, education or institutional land use), residential (housing), agricultural and desert. (a) Per cent cover of patch types. (b) Density of patch types. (c) Mean patch size. (d) Landscape shape index. The section of the gradient covered by more-or-less contiguous built environment (used to approximately delineate the urban complex) is noted in each plot. (Modified from Luck and Wu, 2002.)

landscape-scale studies (e.g. FitzGibbon *et al.*, 2007; Shanahan *et al.*, 2011; Vergnes *et al.*, 2012), mostly based around core tenets of patch size and species–area relationships (see Box 2.4) and patch isolation (see Box 2.5), though the two are sometimes conflated, as ecosystem fragmentation reduces patch size and increases patch isolation concurrently (With and Pavuk, 2011). Consequently, the size and spatial distribution of landscape patches have direct ecological relevance.

Within the landscape, corridors are also considered as landscape elements that may link patches and thereby increase flows, particularly of organisms. Urban corridors are usually linear green spaces connecting other patches of green spaces, and ecologically have been shown to be important for metapopulation persistence

Box 2.4 Patch size and the species–area relationship

The species–area relationship (SAR) is fundamental to ecological theory. For many decades, ecologists have recognised that the size of a discrete area will directly affect the number of species found in it. Early studies based on oceanic islands determined that larger islands maintained more diverse assemblages than smaller islands – this was to prove a central tenet to the Equilibrium Theory of Island Biogeography (MacArthur and Wilson, 1967), which in turn inspired early landscape ecological theory and much nature reserve design. The principle was extrapolated from oceanic islands to ecosystem patches within the landscape, with studies confirming that large patches maintained higher biodiversity, but also that some species required patches of a particular size in order to persist over time. This helped to establish the negative ecological impacts of ecosystem fragmentation – as large patches were reduced to smaller patches, the species diversity decreased accordingly, often following a time lag.

This trend applies to many different species and ecosystem types, though the specific relationship observed depends on the species, ecosystem, wider region,

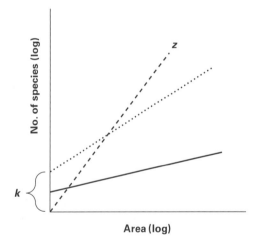

Figure 2.5 The species–area relationship (SAR), demonstrating a positive increase in species number with increasing patch areas under three different scenarios. The constant *k* indicates the regional species richness, indicating that in very biodiverse environments, even small patches are likely to maintain a certain level of diversity. The constant *z* represents the slope of the relationship, showing that the increase in diversity as patch area increases will vary from region to region. The dotted line indicates a scenario in which even small patches will maintain a high level of richness, which will increase notably as patch size increases. The dashed line indicates a scenario in which small patches may maintain few species, but a rapid increase in diversity is found even with modest increases in patch size. The solid line indicates a scenario wherein small patches will maintain some species, but the increase in diversity with increasing patch size is relatively modest. These are just hypothetical examples of the ways in which the SAR may vary from case to case. (For wider discussion, see Spiller and Schoener, 2009.)

and the spatial scale under consideration. The relationship is defined by the equation $S = kA^z$, where S = number of species, A = patch area, and k and z are fitted constants that vary from case to case, and essentially represent the slope of the relationship (z) and the regional species pool (k); see Figure 2.5. Values are often log-transformed. Many explanations for the SAR have been suggested, with the most common ones including (1) larger and more stable species populations being found within larger patches, decreasing the likelihood of localised extinction; (2) increased habitat heterogeneity and therefore niche availability within larger patches; (3) reduced disturbances in larger patches, thereby creating more stable species populations; and (4) increased chance of speciation (the emergence of new species) in larger patches, though this mainly relates to very large ecosystems (see, e.g. Turner and Tjorve, 2005; Spiller and Schoener, 2009).

Box 2.5 Patch isolation

Patch isolation is essentially a measure of how far apart different patches are, which will influence the capacity for exchanges of organisms and materials between them. This was another core tenet of the Equilibrium Theory of Island Biogeography (MacArthur and Wilson, 1967), with the observation that islands that were further from mainland areas experienced decreased colonisation compared to those that were nearby. Extrapolation to terrestrial landscapes helped to inform landscape ecology, with the recognition that ecosystem fragmentation made the movement of species (in particular) between patches more difficult. Isolation is, however, a relative concept, with the functional level of isolation varying between different ecosystems (and the intervening matrix) and individual species. Highly mobile or easily dispersed species may have little difficulty in crossing the distance between patches, while for some species even relatively minor patch isolation may prove problematic (Prevedello and Vieira, 2010). In urban areas, the urban matrix may be relatively hostile as a form of habitat, but may not represent a barrier to movement in some cases. In general, a more connected landscape will have fewer gaps between the ecosystem patches under consideration (because urban areas are of course highly connected with respect to the built environment, for human use, but less so for green space and non-human species), but the amount and significance of patch isolation will vary dependent on the species or process under consideration.

(e.g. FitzGibbon *et al.*, 2007; Magle *et al.*, 2010), and gene flow (e.g. Munshi-South, 2012) in particular, though societal benefits are also apparent (e.g. Ignatieva *et al.*, 2011). Evidence of their importance is not unequivocal, however, and several studies highlight the importance of the wider matrix in influencing connectivity (e.g. Angold *et al.*, 2006; Pino and Marull, 2012). This is mainly because there are particular differences between the structural and functional connectivity of a landscape. Although similar ecosystems may be linked by corridors (whatever their form, e.g. contiguous strips of green space or series of patches) from a human perspective, this does not necessarily mean that flows of organisms or materials are conducted along them. In some cases, corridor characteristics such as width or internal structure may be inappropriate for the movement of species or materials; in others, organisms simply do not disperse in a linear fashion (e.g., seeds dispersed via anemochory). Some organisms simply do not interact with corridors on the (often anthropocentric) scale at which they are designed. This means that although a landscape may be connected structurally, from an anthropogenic perspective, it may not function in a connected way. In contrast, a fragmented landscape may be functionally connected for some organisms or materials, especially if they are highly mobile, such as birds (Tremblay and St. Clair, 2011). This is further complicated by the matrix – although an urban (built) matrix may be relatively inhospitable to movement, some species will be able to move through it relatively easily. Consequently, connectivity varies greatly depending on the scale and the organism/process under scrutiny.

Particularly important for urban ecology and biodiversity is the capacity of groups of patches and corridors to support species metapopulations, which are spatially discrete but interacting populations of a species (Figure 2.6), and thereby determining their long-term fate in the region. The mosaic of different patch types within the urban landscape creates a fragmented habitat template that may support species

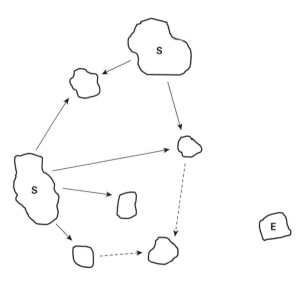

Figure 2.6 An example of metapopulation interactions. The larger patches marked by an 'S' act as source patches wherein reproduction exceeds mortality. As a result, individuals from source populations disperse to the other patches (dispersal represented by solid lines), which may be sink patches where mortality exceeds reproduction. This repeated colonisation from the source patches, along with less reliable 'stepping stone' dispersal between sink patches (represented by dashed lines), helps to keep the metapopulation relatively stable and prevent the population in any one patch becoming extinct. This is highlighted in the patch marked 'E', which is too far away to receive colonising individuals and therefore cannot persist within the metapopulation.

populations as long as interchange of individuals between suitable patches is possible to maintain populations and genetic stability. Connectivity between 'source' patches and other patches is particularly important in maintaining populations and preventing localised extinction (Figure 2.6). The key characteristics of the mosaic that influence species movement/dispersal will vary according to particular urban systems and species, and there are no universally accepted metrics to measure functional connectivity (Magle *et al.*, 2009). Most studies focus on some quantification of patch size, abundance and density, along with distance between patches (as a measure of patch isolation, see Box 2.5).

The reality is often more complex than such metrics can predict, but they have utility for quantifying species patterns and processes. Magle *et al.* (2009) found that of a range of varying metrics designed to measure connectivity, patch size and matrix permeability (i.e. the ability of a given species to cross varying barriers within the matrix, such as roads, or indeed the surface cover of the matrix itself) were the best predictors of prairie dog (*Cynomys ludovicianus*) distributions and dynamics in Denver, Colorado (USA). Mörtberg (2001) found that various *Parus* spp. required woodland patches of >30 ha and ideally >200 ha, alongside a relatively high abundance of urban woodland in the landscape, in order to support populations. Distance between patches was also found to determine patterns of distribution. Several butterfly species examined by Wood and Pullin (2002) around Birmingham (UK) were found to be limited more by habitat availability than distance between patches. These studies help to illustrate that patch size and connectivity are important aspects of spatial urban ecology.

2.5.2 Landscape heterogeneity

All landscapes display some level of heterogeneity, which is essentially the variety of ecosystems or habitats found within a defined spatial area. Urban landscapes are particularly heterogeneous, as their complex and fragmented structure means that many different ecosystem or habitat types may be found in close proximity. This may partly explain some high measures of biodiversity in urban areas, as high heterogeneity results in an abundance of different habitat types and niches for species to exploit within a small spatial area (see Chapter 6). Even small patches will maintain some level of species diversity, particularly of common or abundant species, and highly fragmented and heterogeneous landscapes will have an abundance of patch and corridor 'edges', which may have elevated species diversity (Box 2.3). In natural ecosystems, heterogeneity is often driven by disturbance. Temporally, there is a relation between frequency, intensity or time since disturbance and biodiversity, with diversity generally being highest at moderate levels of disturbance (Figure 2.7). This is essentially because moderate disturbance helps to prevent competitively superior species from becoming dominant and excluding other species, and thereby allows a wider range of species to persist in the ecosystem. The disturbance becomes a resource that species can exploit. Too much disturbance creates a situation in which only very short-lived and prolific species can persist, while too little disturbance leads to species exclusion and dominance of relatively few superior competitors as succession takes place (see Section 6.5).

The spatial extension of this is the disturbance heterogeneity hypothesis, which suggests that while a spatially extensive disturbance (such as a large wildfire) may create

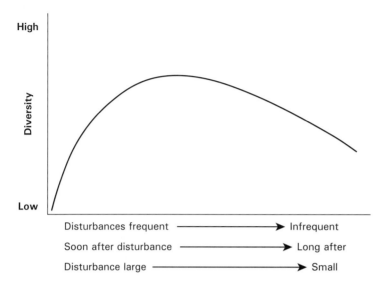

Figure 2.7 The Intermediate Disturbance Hypothesis, illustrating that biodiversity is predicted to be at its highest where moderate levels of disturbance (in terms of severity of disturbance, frequency of disturbance or time since disturbance) are found. Too much disturbance favours those species that are capable of tolerating disturbed conditions, while too little leads to competitive exclusion. A moderate level of disturbance allows species from a range of niches to persist, creating a more diverse assemblage. (Reproduced from Connell, 1978, with permission.)

a relatively homogeneous patch within the landscape, with a similar species assemblage developing in a similar way at a similar rate, different levels of disturbance within a spatially contiguous zone will create many varied opportunities for different species or assemblages to coexist. This hypothesis may be extended to urban landscapes. The built environment may represent the most intense and extensive form of disturbance (though is not entirely hostile to biodiversity; see Chapter 5), while remnant woodland may be at the less-disturbed end of the range. Urban parklands, gardens and regenerating brownfield sites all lie somewhere in between, creating a spectrum of disturbance regimes (for example in terms of time since disturbance, frequency and type of management) that is spatially heterogeneous.

2.5.3 Difficulties with generalisations on landscape structure

However, relating these landscape patterns to ecological patterns and processes may not be as straightforward as it seems. Though we may perceive urban landscapes as highly fragmented, the high density of small patches of habitat in urban complexes (e.g. gardens, lawns, parks, trees) may mean that it functions more as a spatially connected but heterogeneous habitat with large but diffuse (rather than abrupt or distinct) patches and gradients, which will vary with the level of urbanisation and the density of both built structures and vegetated habitat. The practical extent of fragmentation in an urban landscape will also vary depending on the scale and species under consideration (McIntyre and Hobbs, 1999; Savard *et al.*, 2000), further making patch–corridor–matrix generalisations problematic. This may account for the lack of evidence supporting the role of urban green corridors in maintaining some species (Angold *et al.*, 2006), and the high species diversity of some urban regions at the land-scape scale, despite an abundance of 'hostile' matrix (Rebele, 1994; Alvey, 2006; Breuste *et al.*, 2008). The patch–corridor–matrix model is therefore an anthropogenic oversimplification as applied to urban landscape structure, but currently remains the most effective conceptual tool for characterising urban environments (e.g. Herold *et al.*, 2005; Cadenasso *et al.*, 2007).

Although there is abundant literature examining the landscape factors that may determine the diversity, distribution and dynamics of urban species, useful generalities are difficult to make. It is always important to consider: (1) the nestedness of scale and metrics that are being investigated (e.g. whether it is total patch area or density that is key, or patch heterogeneity); (2) the spatial scale of activity for the organism under investigation; (3) what the functional connectivity of the species might be rather than just structural connectivity – as what we may consider to be a structurally connected landscape (e.g. corridors of green space connecting vegetated patches) may not act as such in reality; and (4) landscape dynamics and change, and how they may also have a driving role in patterns found, alongside structure. Much work remains in determin-ing appropriate spatial scales and metrics to look at species population dynamics and other ecological processes within urban landscapes.

2.5.4 Socio-economic patterns and urban landscapes

That socio-economic and cultural groups are spatially organised in urban regions is clear. Marcuse and van Kempen (2002) and Warren *et al.*, (2010) discuss this organi-sation in more depth, identifying that older, pre-industrial urban regions mainly segre-gated social groups according to economic function, so that particular districts (e.g. livestock industries, markets, metalworking, etc.) were established, and there was a more distinct boundary drawn between public spaces (e.g. markets) and residences. Industrial development created more complex divisions within urban regions, includ-ing extensive expansion of residences away from the urban centre ('suburban' areas; see Box 2.2). More recently post-industrial regions have experienced a desegregation of functional divisions, with a trend for urban centres to become more residential and for pockets of gentrification, or the migration of more affluent residents to poorer areas with a subsequent change in socio-economics and culture, to blur the spatial organisation that had become so distinct in industrial regions.

Within this framework of segregation by economic function, further divisions occur according to: (1) 'hierarchical status', reflecting the socio-political power or ability of certain groups to intentionally determine their location of residence as well as the type of built environment they (and others) inhabit; (2) socio-economic status, reflecting aggregations of income and education that enable certain levels of choice of location and organisation; and (3) culture, reflecting the tendencies of groups to spatially orga-nise based on similarities of language, ethnicity and/or religious beliefs (see, e.g. Novy *et al.*, 2001; Iceland and Scopilliti, 2008; Warren *et al.*, 2010). Certainly such organi-sational tendencies within urban environments have existed for millennia (Gates,

2003), and continue to be prevalent in modern complex urban regions (Kinzig *et al.*, 2005; Forman, 2008; Iceland and Scopilliti, 2008).

There are several aspects of this organisation that have relevance for understanding urban landscapes and their ecology. At the hierarchical or political level, urban political and economic theorists have highlighted that spatial inequalities across the social groups have in most cases been purposely constructed, essentially by those at the top of the social hierarchies and with associated political powers to impose legal restrictions on the development of amenities and facilities, allocate resources for education and employment, and generally disempower certain (mainly marginal) social groups or geographic areas, intentionally or unintentionally (Baeten, 2001; Boone, 2008; Warren *et al.*, 2010). As a consequence, at the broadest spatial scales, the abilities of social groups to choose their living environment is affected by such political decisions before individual actions linked to personal preference, education, income and cultural background come into play. Put simply, the ecological quality of a location, and ability of social groups or individuals in that location to engage with urban ecosystems (for example accessing green spaces or actively learning from or facilitating urban ecology), often depends first on decisions made by social and political elites, and only then on the other elements.

At the broadest spatial scales, this can have an effect on factors like the provision and maintenance of green space, localised environmental stewardship and environmental interest, and may in turn be reflected in the urban landscape (e.g. Ferris *et al.*, 2001; McConnachie and Shackleton, 2010). For example Lubbe *et al.* (2010) found that urban segregation resulting from apartheid policies in South Africa mapped strongly onto plant diversity, with white-dominated areas (with a generally higher socio-economic status) displaying higher plant diversities than black-dominated areas, though the 'white' areas were composed mainly of non-native species, presumably intentionally planted and reflecting the economic capacity to engage with horticulture, with native assemblages being more common in 'black' areas. Similar trends have been identified for green space provision in South African urban regions (McConnachie and Shackleton, 2010).

These inequalities are further reinforced by the tendency of much urban growth, especially in less developed countries, to occur in relatively unplanned developments on the edge of the urban complex. In 2010, almost 828 million people were estimated to live in slums around the world, with the highest levels in sub-Saharan Africa (61.7 per cent of the urban population living in slums) (UN-Habitat, 2011). Such areas often have limited infrastructure and facilities and limited access to parks or other recreational green space, though this does not mean that they are necessarily ecologically impoverished (e.g. Lubbe *et al.*, 2010).

For those with the capacity to choose where they live within the urban region, there are many complex factors that will influence their decision, though the environment can be an important variable (Bark *et al.*, 2009; Jim and Chen, 2009). The preferred choice of many people for less-dense housing and more open environment is reflected in the patterns of exurban expansion found in many urbanising regions, which is often outstripping the pace of 'densification' or 'infilling' (see Section 2.6). The result of this is that some regions may have more spatially explicit disparities between socio-economic status than others, for example those that maintain informal housing created by high levels of immigration (Forman, 2008; UN-Habitat, 2011).

Overall, this trend creates a further broad spatial distinction in socio-economic patterns that may influence, for example, patterns of species distribution and conservation engagement. Slums are not planned and consequently may heavily fragment pre-urban ecosystems, the remnants of which may be further degraded or neglected, with little incorporated or designated green space. Such areas will be densely housed with little infrastructure and little concern for diversity, but may take up limited space. In contrast, more affluent areas (e.g. N America) may be subject to greater dispersion (or 'sprawl') with lower building densities. This may fragment natural habitat, but also support citizens who are more in favour of ecological improvements (Robinson et al., 2005; Troy et al., 2007; Fuller et al., 2008). Some poorer and marginalised communities also live in the inner urban complex, essentially in areas that have not had their infrastructure upgraded or maintained, or have been intentionally left to degrade (e.g. Ferris et al., 2001). Although relatively unmaintained, they usually have some level of initial planning that is different from slums on the urban edge, which have a much more random arrangement and limited governance. Often these slums (whether within or on the edge of the urban complex) will have a cultural component; for example many slums in rapidly expanding urban regions are composed of groups of immigrants that originated in the same exurban or non-urban areas, and so have close cultural and sometimes familial ties.

This spatial and dynamic structuring of socio-economic and cultural groups has several implications for urban ecosystems. In particular there are links between: (1) socio-economic pattern, biodiversity and access to green space, with attendant life quality benefits; (2) perceptions of and responses to urban ecology in various forms, including 'nature', 'wilderness', green space and species; and (3) the potential for engagement with conservation activities. All of these may shape the biophysical urban landscape.

2.6 Dynamics of the urban landscape

The relationships between landscape dynamics and structure are not generally well explored, with most studies focusing on a particular 'snapshot' of spatial organisation (Ramalho and Hobbs, 2012). Understanding changing landscape patterns is important, however. The diffusion–coalescence cycle described by Dietzel et al., (2005) predicts certain patterns of landscape metrics, with distance between urban (i.e. built) patches being highest at the early stages of diffusion, before declining as urbanisation continues and more patches are introduced. As distance between patches drops, patch density increases until the patches begin to merge in the coalescence phase and larger, homogeneous built patches are formed. Metrics such as edge density can be expected to peak as the coalescence phase starts. Dietzel et al. (2005) found some relationships between these patterns and the two phases, though these were quite variable, again reinforcing the complexity of urban dynamics. The spatial patterns observed for Phoenix across the urban gradient (Luck and Wu, 2002) reflect these phases, but only to an extent; outer areas that have yet to experience diffusion have large patches and low densities as expected, but the older and more 'coalesced' urban centre still maintains many small densely arranged patches rather than fewer, consolidated built patches as expected. This may emerge with time, however, as green space is lost and built patches join, indicating that the cycle may not yet be fully complete in the urban centre. Other studies have made similar observations, with concurrent expansion and infilling being observed in several urban regions in the USA from 1992 to 2001, but with a reduction of patch density due to urban infilling being less pronounced than might be expected (York *et al.*, 2011). In many cases, patterns depend on how the different patches are defined and metrics are measured. Regardless, landscape metrics are a useful tool to elucidate patterns and dynamics of urban growth, even if they are usually applied to each region as a standalone case.

The development of such landscape patterns is highly dynamic and in many cases characteristic of non-linear systems, with sudden changes in (for example) patch dynamics over time, as development, redevelopment, reconfiguration and new forms of ecosystem management occur (e.g. Luck and Wu, 2002). In some ways this may echo the 'shifting habitat mosaic' found in many natural ecosystems (see Wimberly, 2006), though the drivers within urban systems are of course quite different and the ecosystem types that shift may not remain in balance – green space that is lost may not return very easily. Although there are attempts to look at life-spans of the built components (buildings for example), in reality many changes to artificial systems are driven by socio-economic factors. This is also true for vegetated patches within the urban complex, though a wide range of seral stages may be observed (see Section 6.5). Kattwinkel *et al.*, (2011) have suggested that a shifting mosaic of green and built space, in particular brownfield sites, may support the widest range of species, with both too much and too little disturbance of green space being detrimental. There have been few studies into seral dynamics of such urban ecosystems, however, nor of the significance of these dynamics for urban biodiversity or ecosystem processes. In all cases, localised, bottom-up interactions have been shown to be important for pattern dynamics (Luck and Wu, 2002; Ramalho and Hobbs, 2012), with broad-scale planning having only a secondary role in landscape pattern emergence and change (see Section 3.2.1). In all cases, the temporal aspect of landscape structure, including dynamics and the influence of time-lags in ecological response to landscape change, needs to be considered (Ramalho and Hobbs, 2012).

Some of the most rapid cycles that can impact on the physical landscape and therefore urban ecology are economic cycles. Economic rise and decline may be reflected in changes in infrastructural development, technological advancement or fixed-term political investment, which may operate at cycles of a few years (e.g. the standard term of a political party or other form of governance) to decades (e.g. with changes in transport or information technology that allow restructuring of where and how people work and commute) (Anas *et al.*, 1998; Warren *et al.*, 2010). These trends may be linked not just to the development and expansion of urban regions, but also to spatial patterns of socio-economic groups, along with gentrification and redevelopment within the urban complex during periods of economic growth and stability, or to patterns of abandonment during periods of economic decline. At the broadest spatial and temporal scales, economic patterns are fundamental in driving the diffusion–coalescence cycles; though the general pattern may remain relatively stable, the temporal nature will vary depending on global and regional economies.

This essentially creates a shifting socio-economic mosaic that overlaps with, shapes and is shaped by the biophysical mosaic. Some sections of the mosaic may change quite rapidly, while others may remain stable for long periods of time. Urban ecosystems will consequently be shaped by socio-economic changes at a range of temporal scales. For example, Hope *et al.*, (2003) suggested that woody species composition and lifespan in Phoenix, Arizona were determined mainly by the choices of current residents (within the last decade) rather than other factors, though the full plant assemblage (i.e. including herbaceous species) would also be shaped by factors such as past land use and history of cultivation or land management. An appreciation of socio-economic dynamics is therefore important for developing a full understanding of the urban ecology of a given region.

2.7 Three-dimensional urban structure

Although much work on qualification of urban form and structure focuses on two dimensions only, the verticality and depth of the urban complex also needs consideration in relation to urban ecology. Urban areas can have extreme contrasts in their vertical dimensions, from sewerage or subterranean transport systems deep underground to soaring skyscrapers hundreds of meters high. In effect, an area of (often relatively flat) land has been converted into a three-dimensionally complex structure that has a dramatically increased surface area, and created a range of ecological niches in the same way as extensive rock or talus features in less modified parts of the earth's surface (e.g. Lundholm and Richardson, 2010). With sufficient understanding and ecological engineering the potential for supporting biodiversity and ecosystem processes on this increased surface area may be substantial – of which, more in Chapters 7 and 8. Ecological characteristics of urban built structures (sewerage systems, walls, roofs, etc.) are discussed in more detail in Chapter 5. Here, general three-dimensional structure is considered in more detail.

Structural heterogeneity and complexity is important in ecosystems for creating a range of niches for species to exploit at a range of scales; from studies showing positive relationships between heterogeneity of tree form and foliage to regional-scale relationships between varied topography and biodiversity, the importance of structural heterogeneity and complexity is acknowledged (Johnson *et al.*, 2003; Kostylev *et al.*, 2005; Reid and Hochuli, 2007; Wacker *et al.*, 2008). Structural heterogeneity and complexity differ slightly though the terms are often used in the same way. Structural heterogeneity is the variety of particular three-dimensional environmental factors (e.g. different vegetation types and morphologies) within a given area, while structural complexity is the 3D surficial structure of biota (rippled bark vs. smooth bark) or abiota (a rocky foreshore vs. a sandy foreshore). Complexity may maximise surface area within a given space and create adjacent microhabitats, while heterogeneity places different resources/niches in close proximity. Both may increase species use and biodiversity. Although much of the 3D complexity in urban environments comes from built surfaces, this does create many different forms of artificial habitat in close proximity, which may act as surrogate habitat (sensu Lundholm and Richardson, 2010) for certain species (see Chapter 7). This may in particular support those species that have been naturally selected to exist within cliff or talus habitats, following the 'urban cliff hypothesis' (see Larson *et al.*, 2004; Francis and Hoggart, 2012), though the significance of this has yet to be determined (see Chapter 6).

Clearly 3D surface varies notably with building type and density (e.g. Grimmond and Oke, 1999), but broad-scale quantification of the 3D structure of urban complexes is difficult, and has until very recently depended upon simple maps and scale wooden models that have been used to aid urban planning (e.g. Shiode, 2001). Recent technological advances have allowed more detailed modelling and visualisation using a combination of remote sensing and geographical information systems (GIS) analysis tied to, for example, detailed volumetric computer-aided design (CAD) models, but this has not been applied to ecological studies of the 3D urban complex or at very detailed scales beyond individual buildings. Instead, generalisations such as Darlington's (1981) observation that approximately 1 ha of wall surface exists for each 10 ha of urban land surface within the UK are used to consider habitat availability of the built landscape, while other studies have looked at structural complexity or heterogeneity of green space (Young and Jarvis, 2001).

Three-dimensional structure may be particularly important for providing habitat in individual ecosystem patches. Fine-scale structural features that enhance the 3D heterogeneity and complexity of individual patches have been shown to be important for biodiversity, by providing niches for species to exploit that may otherwise not be present (Young and Jarvis, 2001), though the ecological value of such structures necessarily varies. For example, fine-scale structural heterogeneity, such as woody vegetation within patches, has been shown to be important for determining passerine bird species richness and persistence within urban parks (Fernández-Juricic, 2004). Structural elements may be broadly categorised as woodland (trees etc.), grassland (such as tall herbaceous vegetation), artificial (walls etc.) and water elements. The value of these elements as particular ecosystems is discussed further in Chapters 4 and 5, but their spatial organisation and contribution to the 3D urban complex is important to consider. Woodland elements are usually found in the form of individual or small groups of trees that are remnants from pre-development ecosystems, spontaneous regeneration, or (particularly in paved areas) have been purposefully planted. Single trees and hedges are particularly common in gardens, while rows of trees are most commonly found bordering roads or canals (Young and Jarvis, 2001; Chapters 4 and 5). Dead wood can be relatively uncommon due to removal in urban areas (e.g. Carpaneto *et al.*, 2010; Dawe, 2010), and is most frequently found in gardens or alongside railway or canal corridors. Overall, residential areas tend to contain the greatest frequency and abundance of woodland elements due to the high incidence of gardens and parks (Young and Jarvis, 2001; Tratalos *et al.*, 2007b).

Of grassland elements, stands of tall herbs and grasses are also mainly associated with canals and railways (and to some extent gardens and brownfield sites), while unmanaged short grass is often found at the borders of more frequently used infrastructure, such as paths and roadsides. Lawns or ornamental grasslands are of course ubiquitous and mainly associated with residential areas (Chapter 4; Young and Jarvis, 2001; Robbins and Birkenholtz, 2003). Although they may not have a dramatic vertical structure, grassland elements nevertheless offer near-surface structural heterogeneity and provide niches that may not frequently be found in built environments.

Walls and other artificial structures are of course the archetypal 3D elements of urban areas, and are covered in more detail in Chapter 5. In terms of fine-scale structures considered here, artificial elements are mainly free-standing walls that are remnants from older buildings, boundary walls, small sections of ruins, or rubble piles. These can be surprisingly biologically diverse structural elements (notably for plants), particularly those walls or rough structures that are of considerable age (Young and Jarvis, 2001; Francis, 2011; Lundholm, 2011; Chapter 5).

Fine-scale water elements mainly consist of ponds and drainage ditches; again, these are often associated with gardens, parks or golf courses, but can represent very useful habitat, in particular supporting aquatic species that do not have a great deal of available resources in many urban areas (e.g. Gledhill *et al.*, 2008; Foley *et al.*, 2012; Chapter 5).

Interestingly, such structural elements do not necessarily show a strong positive correlation to patch area, as might be expected; in other words, a small residential area or park may contain just as many structural elements on average as a smaller one. Young and Jarvis (2001), in their survey of urban Wolverhampton (UK) found that some of the greatest numbers of structural elements were found in smaller patches of green space (<10 ha). The use of structural elements to provide fine-scale habitat for both individual species and biodiversity in general is considered more in Chapters 7 and 8.

The urban complex is not three-dimensional only above the ground surface, but also below. Some global cities have extensive underground structures, including underground train lines, etc., as well as the vast network of sewerage and other forms of drainage that urban areas require. Many urban rivers are buried underground and incorporated into sewerage systems (Elmore and Kaushal, 2008). These may create particular niches for species that thrive in subterranean ecosystems. The ecological significance of these is discussed in more detail in Chapters 4 and 5. Subterranean flows are important for pollution removal and urban metabolism, discussed more in Chapter 3.

Understanding the 3D urban complex is important for appreciating ecology and biodiversity, especially as the vertical dimension is the one most likely to expand in some areas, and offers ecological and environmental engineering possibilities for more sustainable development (see Chapters 7 and 8).

2.8 Chapter summary

- Urban regions may exhibit a range of forms, including 'concentric rings', 'dispersed-sites', 'transportation-corridors' and 'satellite-cities', though these are only generalisations. Geographical factors such as rivers, lakes and mountains will further shape urban forms.
- Urban growth usually occurs in phases of expansion and infilling that occur at the same time, but with dominance of the different phases varying temporally. This is often termed 'diffusion' and 'coalescence'.
- The urban complex is usually considered ecologically from a landscape perspective, using the patch–corridor–matrix model.
- The landscape of the urban complex is fragmented and heterogeneous, though distinctive gradients of spatial organisation may be found from edge to centre of the complex. These gradients are usually reflected in landscape metrics, which show a decrease in patch area and an increase in patch density (i.e. greater fragmentation) with proximity to centre. The centre itself may not appear highly fragmented due to the abundance of one land surface type – the built environment.
- Landscape form and structure have an influence on ecological patterns and processes, with fragmentation often reducing habitat area and connectivity, and consequently isolating species populations, leading to loss of some species.

Quantification of the impacts of fragmentation on urban ecosystems can be difficult.

- Urban areas are three-dimensionally complex and heterogeneous, which can create abundant niches for some species. Most of the three-dimensional urban area is made from the built environment, and may have limited use as habitat; maintaining structural complexity and heterogeneity is most important for green space.

2.9 Discussion questions

1   Does your urban region fit into a distinct typology? What factors might have been most important in shaping the region's form? What geographic constraints exist?
2   How abundant is green space in your urban region? If you can view satellite imagery online, is there more or less than you supposed? How is it organised? Does a green belt exist? How close is your home to an area of green space?
3   What are the relative merits of compact or expansive growth of the urban complex?
4   In basic terms, how might structural and functional connectivity vary between: (a) an earthworm; (b) a flying beetle; (c) a small rodent; (d) a bird; and (e) a large mammal?
5   Consider the structural heterogeneity of some different urban ecosystems (e.g. gardens, parkland, brownfield sites, buildings). Which are likely to have higher structural heterogeneity and complexity?

Chapter 3

The urban ecosystem

An overview

3.1 Introduction

This chapter summarises what is known about the ecology and environment of urban regions as ecosystems. Studies of the urban region ecosystem have taken a systems science approach to looking at how such areas function, including how their feedbacks and dynamics influence broad-scale biodiversity, biogeochemistry, climate, hydrology, pollution, flows of materials and energy, and system resilience (Box 3.1; Odum, 1983; Pickett *et al.*, 2001). From this perspective a more 'outwards-facing' approach can be adopted allowing for investigations into interactions between the urban region and other ecosystems outside its broad spatial delineation (for example quantification of the ecological footprint; Wackernagel and Rees, 1996). Indeed, Pickett *et al.* (2011) note that this more comprehensive consideration of urban regions is the most useful for developing a scientific understanding of urban ecology and how urban environments can be managed, though there has been limited research in this area compared to some other more localised aspects of urban ecology.

First the chapter considers some key aspects of urban areas as 'systems', covering some central concepts related to this, before going on to look at the general characteristics of some of the key components and processes of the 'urban ecosystem', including biogeochemistry, climate, hydrology, soils, pollution, and biodiversity.

3.2 Urban regions as complex adaptive systems

Ecosystems can be viewed as complex adaptive systems (Levin, 1998). This is because they are composed of interacting components which have the capacity to change in response to both internal and external environmental influences (e.g. Levin, 1998; Currie, 2011; Cook *et al.*, 2012). They can generally be characterised by:

1   multiple components that interact across a range of scales;
2   non-linear patterns and processes;
3   existence in a state apart from equilibrium that requires a flow of energy for maintenance;
4   self-organised properties;
5   the capacity of both components and the system to 'learn' or change in response to environmental factors (this can be both internally and externally driven change).

Box 3.1 Urban resilience

Any complex adaptive system will have some level of resilience to disturbance based on the capacity of the multiple interacting components to differentially respond to stresses and maintain structures, functions and processes. Discussion of urban 'resilience' is very common, and Leichenko (2011) notes that these usually focus on ecology, risks relating to hazards and disasters, economy, and governmental or institutional capacity to improve resilience. Most work has focused on the social aspects of resilience, essentially looking at how social urban systems may be better enabled to withstand stresses without highly detrimental impact, and how quickly the systems may return to normal after stress. However, this interpretation, though valid, reflects the social sciences; the original concept of resilience within ecology (Holling, 1973) did not indicate the capacity of a system to recover from shock and the speed with which it recovers – this is more 'stability' or 'resistance'. Resilience was rather the amount of disturbance a system could take before it would move from one 'stable domain', or ecosystem state, to another. The terms 'engineering resilience' and 'ecological resilience' were subsequently applied to distinguish between the two; keep striking a hammer and it will absorb the disturbances ('engineering resistance') until it breaks, when it no longer functions as a tool. Disturb an ecosystem sufficiently and it will change state, but into a new functional ecosystem. Most urban interpretations of resilience, including ecological resilience, focus more on this idea of resistance, i.e. enabling the ecology of an urban system to cope with, e.g. climate change (Leichenko, 2011). This is entirely valid and an important consideration, particularly alongside coupled ecological and socio-ecological systems; but perhaps the crucial factor for an urban system truly is resilience, i.e. how much disturbance or stress is required to change an urban ecosystem from its main purpose (a mainly artificial system constructed to allow humans to live in relatively dense populations) to something else. Greater consideration is needed of interpretations, measures and significance of both resistance and resilience within urban ecosystems.

Urban regions display all of these characteristics. For example, multiple components of biota and abiota interact across a range of temporal and spatial scales. Non-linearity is demonstrated by hydrological response to different levels of impervious surface cover, or biodiversity in relation to patch size. Urban regions also have non-equilibrium states maintained by anthropogenic energy subsidies in the form of fossil fuels. Self-organisation can be found in the emergent patterns of buildings and infrastructure and adaptive changes occur constantly in the form of succession, species interactions, natural selection, and other processes. Perhaps the most interesting aspects of these within the urban context are self-organisation and urban metabolism, the latter in particular reflecting non-equilibrium states.

3.2.1 Self-organisation within the urban ecosystem

Self-organisation is common in nature and indeed integral to biology and ecology (Odum, 1988). Many ecosystems display evidence of self-organisation via reciprocal interactions between biota and abiota. A good example of this is that organisms can 'engineer' ecosystems resulting in the emergence of broad-scale patterns (e.g. Rietkerk and van de Koppel, 2008). There has been much recent work on the emergence of patterns and processes in urban environments, and it is clear that many elements of urban ecosystems display evidence of self-organisation (e.g. White and Engelen, 1993; Jiang, 2007). This may seem somewhat counter-intuitive in that urban systems are highly 'artificial' and appear to have been rigorously planned or designed by humans. But the reality is that much urban growth has historically been relatively unregulated, and development was driven perhaps by human instinct and localised necessity or requirements rather than design. Given this, many emergent patterns and processes are not intentionally planned, but have resulted from a multitude of localised human decisions that tended to have similar objectives.

This is reflected in patterns of landscape metrics observed for urban regions studied (see Chapter 2) in terms of patch size, density and so on – the trends observed, such as a peak in patch density and a nadir in patch size during a 'diffusion' cycle, and spatial metrics of such gradients across a region, are common or even universal, but have not been specifically planned.

As an example of this phenomenon, Jiang (2007) highlights that urban street networks universally display an 80–20 relationship where approximately 80 per cent of roads have lengths and degrees of connectivity with other roads that are less than the mean network value, while 20 per cent have lengths and connectivity much greater than average. 'Backbone' routes which support the network represent less than 1 per cent of all streets. This essentially is the recognition that urban areas have key roads running through them that are primary routes for connecting the wider road network, and which support the majority of traffic flow (80 per cent of traffic being conducted along 20 per cent of the roads). Likewise, the very few key roads (<1 per cent) act as 'cognitive maps' of the urban region, via which people create their spatial awareness of the area for navigation. While this is an interesting pattern in itself, it is the fact that the approximate 80–20 ratio is widespread that suggests a self-organising pattern. Urban designers did not plan a network with such a ratio – indeed, in those areas that have very regulated grid systems, the pattern is much weaker or not found at all (Jiang, 2007).

Self-organisation is clearly fundamental to our understanding and management of urban regions as ecosystems. As Kühnert *et al.* (2006) note, most urban planning is based in reality on changing or modifying what already exists, and the patterns that have already emerged via unclear processes, rather than developing grand designs from scratch.

3.2.2 Urban metabolism and wider impacts

Considering urban regions as acting like heterotrophic organisms, taking in resources from outside and producing waste products that are expelled, is a useful analogy for the way these ecosystems function in relation to their surrounding environment (though is not always considered appropriate; Grimm *et al.*, 2008). No ecosystem is isolated, and all are nested within other systems as noted in Section 1.3. Urban regions in particular are major consumers of environmental resources (food, fuels, building materials, water and so on) at local to global scales. Urban regions also produce abundant waste in the form of pollutants, which must be processed by other ecosystems within the biosphere (Folke *et al.*, 1997). Studies that have attempted to quantify rates of consumption and waste or the overall 'urban metabolism' have focused on two different approaches: energy equivalents and mass fluxes (see Kennedy *et al.*, 2011).

Determination of energy equivalents is an approach that follows Odum (1988), and uses the unit of 'emergy'. Emergy is simply the available energy used up in the creation of a product or completion of a process. The baseline for emergy is usually the amount of solar energy required to achieve a product or service, and is expressed in the unit of solar emjoules (seJ). This approach allows for a standardisation across all systems and contexts, as solar energy is the fundamental energy source for the entire biosphere. Unfortunately, there have been relatively few studies to date that have attempted to quantify urban products and flows using this system (see Huang and Hsu, 2003; Zhang *et al.*, 2009; Jiang and Chen, 2011).

Quantification of fluxes is usually based on more common units, for example tonnes of carbon dioxide equivalent (t CO_2 e) when referring to carbon emissions of an urban area (see Kennedy *et al.*, 2011 and examples therein). Quantification and mapping of such fluxes is useful in determining what the 'ecological footprint' of an urban region might be. In a landmark study, Folke *et al.* (1997) calculated the ecological footprint or 'ecosystem appropriation' of 29 urban regions in Baltic Europe. This research showed that on average 1 km^2 of urban area utilised 18 km^2 of forest, 50 km^2 of arable land and 133 km^2 of marine ecosystems in order to supply fuel and food resources. Based on processing of excess nitrogen and phosphorous, and carbon sequestration, processing of urban wastes required 11–30 km^2 of arable land, 48 km^2 of inland water, 30–75 km^2 of wetlands and 354–870 km^2 of forest. This was a phenomenal 565–1130 times greater areal extent than the 1 km^2 urban area itself. Other studies have made similar findings as well. Vancouver, Canada, utilises 180 times more land than its own size to provide the resources it consumes (Pickett *et al.*, 2008), while London (UK) is estimated to require more than 300 times its spatial extent (Greater London Authority, 2003). Even in 1997, Folke *et al.* (1997) concluded that 20 per cent (1.1 billion) of the global population, living in 744 large urban regions, produced over 10 per cent more CO_2 emissions than the total storage capacity of global forest ecosystems. Likewise, Odum (1997) demonstrates that heavily urbanised regions may consume 100,000–300,000 kcal m^{-2} yr^{-1}, compared to the 1,000–10,000 kcal m^{-2} yr^{-1} of more natural ecosystems; over an order of magnitude greater. It is clear that urban ecosystems have a dramatic impact on their environment, and their impact can extend from local to global scales, with notable spatial heterogeneity (Luck *et al.*, 2001). This is mainly due to industrial development and the availability of fossil fuels to provide sufficient energy to power the systems. Natural ecosystems are of course maintained by energy inputs from the sun and are therefore more or less sustainable, in a dynamic balance with the available energy. Urban regions can only exist due to heavy injections of fossil fuels that enable their intensive energy requirements to be met. Making urban regions more energy efficient and self-supporting, and thereby reducing their ecological footprint, is a major challenge for sustainable development, but one that must be met as urban regions continue to grow globally.

3.3 Urban ecosystem services

All ecosystems, including urban ecosystems, provide 'the services, goods and benefits gained from the environment that benefit humans' (Lundy and Wade, 2011, p. 654). Considering these as 'services', rather than simply environmental processes and functions, provides an intrinsic human dimension. As long as at least one person benefits from a given process or function it can be qualified as an ecosystem service (Karieva *et al.*, 2011). With a sufficiently loose interpretation, pretty much any environmental process on Earth may indirectly benefit humans, but certainly urban ecosystems very clearly and directly support human populations. Because of this, urban ecosystem services are important to consider. The human element also helps to emphasise that ecosystem services are an *applied* concept for the present day. Without such services, our civilisation and indeed our species, could not survive.

Ecosystem services is an important concept allowing for the quantification of environmental processes so that they may be incorporated into management of the environment (Kremen and Otsfeld, 2005; Carpenter *et al.*, 2009). Whether this is entirely ethical or not (Goulder and Kennedy, 1997), it is perhaps the best language for cutting across scientific disciplines and allowing environmental managers to appropriately consider the implications of the loss of services and ways in which losses may be mitigated or reversed. The services are usually categorised into provisioning, regulating, cultural and supporting services. There are three main providers of services within the urban ecosystem; vegetation ('green infrastructure'), soils ('brown infrastructure') and water ('blue infrastructure'). As with any complex adaptive system the components and their processes are interrelated, but key services provided by these components in urban ecosystems are as follows.

3.3.1 Provisioning services

These cover the provision of products from ecosystems, including water, food, fuel and medicine, among other things. This is perhaps the most well-documented form of service, and the easiest to quantify economically. Urban vegetation provides fuel (particularly in developing regions) and food from agricultural areas or allotments/gardens (Leake *et al.*, 2009). Aquatic ecosystems may supply much potable water, from aquifers, rivers, reservoirs or directly channelled stormwater. Urban waterways may also contribute to local hydropower schemes (Lundy and Wade, 2011), as well as providing food sources such as fish and crustaceans.

3.3.2 Regulating services

Regulating services are those that regulate processes or functions that may have detrimental human impacts, such as flooding, pollution, soil erosion and climate change. Vegetation, soils and water all help to sequester carbon, store, break down, disperse and dilute pollutants, provide habitat to prevent biodiversity loss and so on. Vegetation and water can be particularly useful in mitigating the urban heat island effect and reducing noise and light pollution (Carpenter *et al.*, 2009; Lundy and Wade, 2011). Even relatively 'low-value' urban habitats such as lawns can fix relatively large amounts of carbon due to their artificially high productivity levels and longer growth seasons, though of course mowing and biomass removal may limit this beneficial aspect to an extent (Pouyat et al., 2010). Regulating services are harder to quantify but are particularly important within the context of mitigating anthropogenic environmental impacts.

3.3.3 Cultural services

Cultural services are the spiritual, aesthetic, recreational and educational benefits that people obtain from ecosystems. These are wide-ranging and important (see Clayton and Myers, 2009), and may have benefits from improving quality of life to economic effects such as improved house values (Jim and Chen, 2009). Vegetation and water have well-established psychological and health benefits (e.g. Clayton and Myers, 2009), and both green and blue infrastructure feature heavily in many recreational activities. Quantifying cultural services can be difficult and quite subjective, particularly as some elements (spirituality, for example) do not lend themselves well to economic valuation (Daily et al., 2009).

3.3.4 Supporting services

Supporting services are those that allow the other services to exist, by underpinning essential ecosystem processes. Examples would include primary production and the creation of biomass, water and nutrient cycling, habitat provision, soil formation, and pollination of plant species. They are long-term processes operating in the background and are perhaps the least understood services and among the most difficult to quantify. Soils, water and vegetation all contribute to these services and are fundamental to a fully functioning ecosystem. They are especially important in urban ecosystems where the quality of all forms of ecosystem service may be compromised.

3.3.5 Disservices

Disservices are ecosystem processes or functions that have a negative effect on human well-being (Escobedo et al., 2011). It is also important to consider the role of disservices, particularly in urban areas with high population densities, where such disservices may have a notable impact (Lyytimäki and Sipilä, 2009). For example, the loss of wetland ecosystems is common within urban areas (see Forman, 2008; Chapter 2), but rehabilitation or ecological improvement attempts that increase wetland areas may have unforeseen consequences such as an increase in insect pests such as mosquitoes, which may in turn have human health implications (Dunn, 2010). Likewise, recreational and health benefits provided by urban water features may be compromised by increased exposure to pollutants or waterborne diseases. Street trees may have many uses, but can also be an annoying source of allergens, a consumer of water supplies, or a hazard in high winds (Escobedo et al., 2011). When regulating services are in place within a functionally integral ecosystem such disservices may be reduced, but their presence and management implications will always need to be recognised for a balanced understanding of urban ecosystem services.

This chapter now focuses on some key environmental characteristics and conditions typical of the 'urban ecosystem'.

3.4 Urban biogeochemistry

Biogeochemical cycles within ecosystems are mainly considered within the context of key elements for biota, such as nitrogen, phosphorus and carbon (e.g. Grimm *et al.*, 2010). Within the urban region fluxes occur between land, atmosphere and water, and within particular terrestrial and aquatic components (Kaye *et al.*, 2006). Fluxes to and from the atmosphere are perhaps the most well-documented because they often relate to urban pollution (see below), and include the addition of carbon dioxide (CO_2), nitrogen oxides (NO_x) and methane (CH_4) from combustion associated with transport and industry, as well as soil emissions often associated with soil organisms (Kaye *et al.*, 2004). Atmospheric deposition of both nitrogen and carbon is relatively high in urban areas, often associated with high levels of dust in the atmosphere that may accumulate particles of these elements. Uptake of these elements also occurs in urban ecosystems, with many forms of vegetation proving important for fixation of CO_2, N_2O and CH_4, though the capacity of urban vegetation to store such elements is often compromised compared to pre-urban or non-urban ecosystems (e.g. Kaye *et al.*, 2004; Groffman and Pouyat, 2009).

Within particular terrestrial urban locations, concentrations and fluxes of carbon, nitrogen and phosphorus will reflect not just rates of emission, deposition and fixation, but also previous land use (e.g. agriculture or semi-natural woodland) and time since urbanisation. The history of land use can relate to, for example, the level of organic material found within soils, which in turn will influence carbon, nitrogen and phosphorus levels (Lewis *et al.*, 2006; Kaye *et al.*, 2008). Anthropogenic (artificial) soils for example may have much lower organic carbon content than more natural soils, with implications for carbon cycling (Pouyat *et al.*, 2006). Active management of vegetated areas, particularly gardens and lawns, may also lead to elevated nutrient levels due to the addition of nitrogen- and phosphorus-based fertilisers to aid plant growth (Groffman *et al.*, 2009). These elevated levels of nutritional resources may also help to increase net primary productivity (NPP) and therefore carbon fixation; indeed, some vegetated urban areas (e.g. parks, gardens, golf courses) can have very high NPP, particularly compared to surrounding exurban or non-urban areas (Imhoff *et al.*, 2004; also, e.g. Wu and Bauer, 2012).

Aquatic urban concentrations of nitrogen and phosphorus in particular can also be high due to direct additions from wastewater treatment plants and other effluents or from diffuse sources like road runoff (see Section 3.8) and atmospheric deposition. Although rivers in particular may rapidly export entrained elements from the urban region, high loading rates often counteract these exports. Some nitrogen and phosphorus also filters through to groundwater stores, particularly in humid areas where groundwater levels may be close to the surface; this is particularly common where there are leakages from sewerage infrastructure (Lerner, 2002). Carbon content in urban river systems is often reduced and this can be due to many factors. Specifically, urban rivers can have reduced inputs of coarse organic matter due to reductions in riparian vegetation, while simplified geomorphology and altered flow patterns (see Section 3.6) can result in decreased retention of organic matter (Chadwick *et al.*, 2010).

These examples are indicative of general changes to biogeochemical cycles, which will clearly impact upon ecological assemblages, influencing which species are present and their growth and reproduction.

3.5 Urban climate

The climate of a given area is usually defined in terms of averaged long-term patterns and dynamics of temperature, precipitation and wind, among other things. Although urban regions do not have identical climates, nevertheless they often display similar trends in their temperature and precipitation regimes (e.g. Shochat *et al.*, 2006). The most well-known is the urban heat island effect, which consists of elevated surface temperatures in the urban complex compared to the surrounding exurban or non-urban land, alongside decreased diurnal temperature ranges due to heat being retained by constructed surfaces (Coutts *et al.*, in press). Mostly, elevated temperatures in urban regions are caused by heat emission from residences and industry, the built environment itself and atmospheric pollution. In urban centres, air temperatures have been recorded up to 10°C higher when compared to non-urban areas (e.g. Zipperer *et al.*, 1997). However, average differences in most regions are in the order of 1–2°C (Pickett *et al.*, 2011).The relative differences between the heat island effect in the urban complex and outside also vary seasonally, with differences being more extreme and consistent in winter within temperate climates (Schlünzen *et al.*, 2010).

Other important factors that influence urban heat island effects include the size and density of the urban complex, composition of buildings, the geographical context (e.g. elevation, latitudinal position, proximity to water bodies, regional climate) and the amount of urban vegetation present (e.g. Collier, 2006; Coutts *et al.*, in press). The urban heat island effect has been linked to population size in some urban areas, but a more accurate predictor of the effect and its severity may be the density of the built environment and abundance of impervious surfaces (Klysik and Fortuniak, 1999). Urban vegetation may substantially lower the effect due to its capacity to cool down built surfaces by intercepting sunlight, providing shade and increasing evapotranspiration to increase moisture in the air. Urban water features may also have a beneficial cooling effect (Coutts *et al.*, in press). The desert-based urban region of Phoenix, Arizona (USA) has cooler summer temperatures than the non-urban region due to its water features and green space, for example (Pickett *et al.*, 2011). The spatio-temporal complexities and uncertainties of the urban heat island effect are too elaborate to discuss here, but see Stewart (2011) and Coutts *et al.* (in press) for reviews.

Precipitation patterns in and downwind of urban areas are also influenced by the urban environment (Collier, 2006). This is mainly because the airflows over the urban complex are disrupted via increased surface roughness of buildings, and convective processes linked to the urban heat island effect help to create increased downwind cloud formation and precipitation (Collier, 2006; Schlünzen *et al.*, 2010). The presence of particulates (e.g. pollutants) that act as condensation nuclei in urban areas are also suspected to influence rainfall, but further work is needed to determine the exact importance of this aspect as some aerosols can actually inhibit convection and resultant precipitation (Collier, 2006).

The urban heat island and related climatic changes affect the ecology of urban environments in several ways. Plant and invertebrate species may experience changes in phenology, with plants displaying earlier germination or flowering dates in urban compared to non-urban areas due to earlier springs and longer periods without snow or frost (e.g. Wilby and Perry, 2006; Jochner *et al.*, 2012) and insects having earlier emergence dates (Wilby and Perry, 2006). These effects are not universal, however, because of variations in level of urbanisation and other confounding factors such as altitude (as most urban areas are found at low altitudes compared to some surrounding areas) (Jochner *et al.*, 2012). Animal behaviours can also be affected, with altered migration patterns, increased reproductive season and changes in egg-laying times observed in avifauna (Wilby and Perry, 2006; Chapter 6). Of course, elevated temperatures can also have significant human health impacts, for example increasing heat stress that can lead to higher mortality of vulnerable groups such as the aged, and increased ozone production.

3.6 Urban surface and groundwater hydrology

Hydrological cycles are significantly disrupted by urbanisation. Although this may in part be due to the direct alteration of surface and groundwater flows and storage by regulation and abstraction, the main driver of changed hydrology is the creation of impermeable (or less-permeable) surfaces. Although such materials (such as pavements, roads and roofs) are often artificial and intentionally created, impermeable (or impervious) surfaces also include natural bedrock outcrops and compacted soils (Arnold and Gibbons, 1996). Less-permeable surfaces have existed in various forms for millennia (e.g. farm tracks, non-paved roads), but it is only since the early twentieth century that such surfaces have become much more abundant, mainly due to the success of the motor vehicle and associated infrastructure.

Impervious surfaces vary with land use. Residential areas contain less impervious surface compared to industrial, commercial or shopping centre areas. They tend to vary notably in their density and their proportion of built surface (including paths, roads and houses) to vegetated habitat (such as gardens, parks or other forms of green space), from around 20 per cent impervious cover upwards (Arnold and Gibbons, 1996). Industrial and commercial areas often have abundant built surfaces due to large buildings, car parks and little demand for local green space, and may have over 75 per cent impervious cover. In an early survey, the United States Soil Conservation Service (1975) found that shopping centres maintained the greatest cover (with an average of 95 per cent) due to their huge buildings and extensive parking spaces. Not surprisingly, clear correlations have long been established between population density and impervious surface cover (e.g. Stankowski, 1972).

Impermeable surfaces have a dramatic influence on urban hydrology (Figure 3.1). First, impermeable surfaces prevent infiltration of rainwater into the soil. This results in increased surface runoff which creates problems associated with both flooding (of drains, sewers and rivers) and pollution entrainment to urban waterways. Arnold and Gibbons (1996) note that for a catchment with natural surface (soil and vegetation), approximately 25 per cent of rainfall infiltrates into the upper soil layers, 25 per cent into deeper layers, 40 per cent re-enters the atmosphere via evapotranspiration and 10 per cent is lost to surface runoff. For a catchment with 75–100 per cent impervious surface (such as found on modern roads and car parks), only 10 per cent of rainfall infiltrates to the upper soil, 5 per cent to deeper soils, 30 per cent experiences evapotranspiration and 55 per cent is lost to runoff. However, the effect of impervious surfaces can be moderated by the effective impervious area (EIA) in a catchment. EIA refers to the impervious areas within a catchment that are connected directly to drainage systems, rather than causing overland flows. EIA can be difficult to quantify and

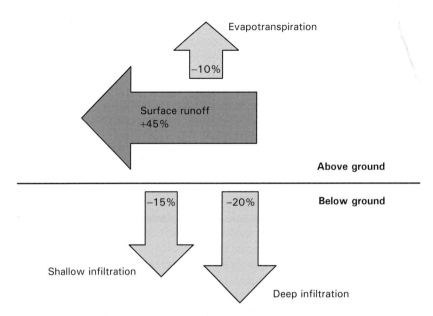

Figure 3.1 Relative changes in the hydrological cycle due to urbanisation, from a natural surface to one with over 75 per cent impervious surface cover. (Based on Arnold and Gibbons, 1996.)

may vary according to level of rainfall and soil saturation; see Jacobson (2011) for further discussion.

Increased surface runoff, especially during heavy rainfall events, causes greater discharge and flashiness of urban rivers, soil erosion and increased sediment loads (Paul and Meyer, 2001; Gurnell *et al.*, 2007). Generally, urban drainages have increased peak discharge (as more water is flowing into the urban river) and reduced time to peak flow (as the water runs directly to the river rather than being incorporated into subsurface flows or groundwater). During dry periods, urban catchments tend to have reduced baseflow, as subsurface flows that compose the majority of baseflow are reduced (Jacobson, 2011). The level of impermeable surface cover needed to significantly change hydrological cycles can be quite low. Estimates vary from location to location and range from 2–20 per cent total impervious area (Paul and Meyer, 2001; Yang *et al.*, 2010).

Altered hydrology in urban systems can result in increased flooding. Many catastrophes that occur each year involve floods, and these are often particularly devastating in urban areas due to their high population densities and the high economic value and density of goods and infrastructure. These factors result in high mortality and costly economic losses when compared to rural areas. In unplanned urban areas in developing countries, where the capacity to respond to disasters is reduced and/or where population pressure causes urbanisation on less suitable ground, impacts of urban flooding can be even more costly. A recent example is the flooding and associated landslides that occurred in the Brazilian state of Rio de Janeiro in 2011, with at least 890 people killed in several different urban areas (Zucco *et al.*, 2011).

A common engineering response to altered urban hydrology is the channelisation of urban rivers to enhance conveyance. Channelisation both increases flow capacity and facilitates the rapid removal of excess water and associated pollutants. However, this type of engineering often causes problems upstream and/or downstream of the urban region. Further, an increase in drainage infrastructure, including river engineering and channelisation, further degrades aquatic and riparian habitats (Groffman *et al.*, 2003). The diversion of water from infiltration to surface runoff also leads to reduced aquifer storage and can lower water tables, leading to both a reduction in available water for essential services and also increased risk of subsidence.

3.7 Urban soils

Soils are the basis of many terrestrial ecosystems, providing a growth substrate for vegetation and supporting a wide range of biota and ecological processes. Urban soils can be highly spatially and temporally heterogeneous and often display characteristics of high levels of disturbance (in terms of frequency, duration and severity) and pollution (discussed below) (Pouyat *et al.*, 2010). Remnant areas may retain soils with relatively natural or semi-natural characteristics (e.g. woodland patches), depending on pre-urban land use (Pouyat *et al.*, 2010). Many urban soils may be formed from anthropogenic sources, for example construction debris, extracted fluvial gravels, solid waste, or natural soils that have been moved from their point of origin and consequently reconfigured. This is in contrast to natural soils that have formed either from bedrock weathering or from the breakdown of materials deposited by natural flows of water, ice or wind.

Key characteristics of soils are their moisture and nutrient content, much of which is determined by soil structure and organic content. Most urban soils have increased compaction due to the weight of overlying or adjacent buildings, and pedestrian or vehicular use. Such compaction reduces the ability of the soil to allow water to percolate and be stored in pores, contributing to the hydrological problems of urban areas noted above. This can be further exacerbated by the tendency of urban soils to form hard surface crusts that further restrict percolation (Craul, 1992) and increase soil evaporation (Pouyat *et al.*, 2010). Pavao-Zuckerman (2008) cautions about over-generalising urban soils because of their high spatial variability. For example, compaction may not be excessive in urban green spaces and may be less severe than in agricultural areas. Edmondson *et al.* (2011) found that urban soil bulk density measurements taken to 14 cm depth were highly variable but overall significantly less compacted than agricultural soils, and were generally similar to many semi-natural vegetated ecosystems, particularly within green space with abundant tree cover (as opposed to, e.g. lawns). In contrast, some areas of urban soil may have elevated moisture content due to leakage of water pipes, which can be extensive (Lerner, 2002) and poor drainage and sewerage infrastructures. Alongside potential pollution impacts, this can lead to localised soil saturation or flooding, and creates a complex mosaic of soil conditions that can affect soil biota and vegetation (Pouyat *et al.*, 2010). These comparisons only consider surface soils in green spaces, however, and soils under impervious surfaces will certainly have altered hydrology.

Many factors affect the quality of urban soils. Reduction in water within urban soil, alongside the localised effects of the urban heat island, means that urban soils often have elevated temperatures. Urban soil pH can be increased by the unintentional addition of (for example) cements (e.g. limestone components) and salts used to de-ice roads. Some urban soils are also likely to have reduced organic matter inputs, thereby disrupting nutrient cycles that are essential for the integral functioning of many soil communities and processes. Despite this, Pouyat *et al.* (2010) note that urban soils remain an important part of the urban ecosystem, and reinforce the consideration of soils as 'brown infrastructure' that in their own way are just as important as 'green infrastructure' (vegetation) in ensuring the functioning of the system and provision of ecosystem services.

Urban soils are frequently disturbed by construction or management. These actions, whether by the addition of pollutants or repeated compaction, can dramatically influence the structure and characteristics of soils to a great depth. For example, compaction may extend to substantial depths beneath buildings and infrastructure, while the construction of subsurface infrastructure may cause instability, compaction or localised pollution. Time since disturbance will also affect soil characteristics. Scharenbroch *et al.* (2005) found that older urban soils (mean age 64 years) in Moscow, Idaho and Pullman, Washington (USA) had higher nutrient and organic content than new urban soils (mean age 9 years). Further, the older soils were closer in typology to the soils of more 'stable' ecosystems suggesting that some level of recovery had taken place in the time since initial disturbance.

The ecology and biodiversity of urban soils deserve particular consideration. Soil biodiversity within urban areas is poorly explored, such that species entirely new to science have been found in urban complexes (Kim and Byrne, 2006; Pouyat *et al.*, 2010). As noted above, soil type and origin, compaction, temperature, moisture, pH and location will affect soil quality and in turn biodiversity (e.g. Schrader and Böning, 2006). Abundance of soil invertebrates in particular seems highly spatially variable, with Santorufo *et al.* (2012) finding abundances of 6000–41,000 individuals m^{-2} of soil in Naples (Italy). Species richness seems less variable, though higher species richness may be found in soils with more organic matter and water, and lower richness in polluted soils (Nahmani and Lavelle, 2002; Santorufo *et al.*, 2012). Common soil taxa such as Acarina, Nematoda, Collembola, Diplopoda and Diptera larvae remain both frequent and abundant in urban soils, though variation in assemblage composition is often found, which may have knock-on effects on biogeochemical cycles. For example, a decrease in collembolans, which seem more sensitive to heavy metal content, may be associated with an increase in nematode numbers (Santorufo *et al.*, 2012). Pavao-Zuckerman and Coleman (2007) looked at how nematode diversity and assemblage composition changed between urban, suburban and rural locations around Asheville, North Carolina. All sites had low elevations, sand loam soils, and hardwood-conifer vegetation communities and were therefore relatively comparable. Soil samples were taken from sites in each location, and nematode fauna identified and quantified. Diversity was not different between locations, while species richness for soils greater than 5 cm deep were found to vary. Specifically, urban locations showed the lowest richness. Notably, predatory and omnivorous nematode abundance was lowest in urban areas. As with aboveground urban habitats, non-native species can be frequent and abundant; for example, no native earthworms were found in the urban complex of New York City by Steinberg *et al.* (1997).

Despite the complex mosaic of soils found in urban areas, soils are similar in many of their symptomatic characteristics between urban complexes, indicating a homogenisation of soils at the global scale (Pouyat *et al.*, 2010). The extent of such variation and its ecological significance remains an important area for investigation.

3.8 Urban pollution

Pollution is a ubiquitous feature of urban environments. It generally peaks temporally at the period of initial and rapid industrialisation and urbanisation before declining as pollution regulation, management and waste-control technologies are applied. Although many developed urban regions are now less polluted than during periods of peak urbanisation, they still remain centres of waste production with both human and ecological impacts from pollution being prevalent. In developing urban regions, pollutant impacts can be severe. Some of the most polluted urban areas in the world include Linfen, China (particulate matter from coal burning), Tianying, China (from heavy metal industries), Sukinda, India (from chromite ore mining and processing), Vapi, India (from industrial chemicals and heavy metals) and La Oroya, Peru (from heavy metal industries) (see Blacksmith Institute, 2007). In each of these examples, thousands to millions of people are affected by the pollution. Sadly, there is limited detailed investigation of ecological impacts in these heavily polluted systems and much more research is needed.

The main forms of pollution found in urban environments are described below.

3.8.1 Atmospheric pollution

Atmospheric pollution and air quality is a major issue within urban regions, for both human health and ecological quality. Air quality varies according to urban population size and density, transportation and industrial activities, and level of development. Many urban regions in industrialised nations may have relatively clean air compared to historical early-industry conditions, while more developing areas may still be highly polluted. Major atmospheric pollutants include particulate matter (PM) that is less than 40 μm in aerodynamic diameter, ozone (O_3), sulphur dioxide (SO_2), nitrogen oxides (NO_x), carbon monoxide (CO), volatile organic compounds (VOCs) and polycyclic aromatic hydrocarbons (PAHs). All of these pollutants are of some significance, but the key ones for human health are perhaps SO_2 and PM (mainly PM_{10} and $PM_{2.5}$), both of which have been linked to premature mortality (Molina and Molina, 2004). Visibility is mainly compromised by $PM_{2.5}$ concentrations, while vegetation is mainly affected by O_3, which causes photochemical oxidant damage, such as leaf lesions. SO_2 and NO_x are central to environmental acidification, which can have significant detrimental impacts on both aquatic and terrestrial ecosystems.

Pollutant distributions and concentrations are determined by factors operating at a range of scales, from the location of sources to the spatial configuration and density of the built environment that affects air flows and pollutant residence times. Sometimes localised spikes of pollution can occur for example in urban canyons, which can often trap pollutants near ground level due to alterations in wind dynamics (e.g. Richmond-Bryant *et al.*, 2011). Such high levels of pollution may contribute to increased incidence of respiratory disease or allergic responses, both of which are common in urban regions compared to elsewhere (Cohen *et al.*, 2004). This is an important and complex issue, and understanding the effects of urban form and structure remains a research priority.

These pollutants are also exported from the urban system and can have regional and global effects. Urban atmospheric pollution may contribute to global warming, especially given the capacity for pollutants to be transported in the atmosphere over long distances. At regional scales, pollutants like O_3 may scatter or absorb solar radiation, both cooling the land surface and reducing sunshine duration. PM may alter precipitation patterns as noted above, acting as condensation nuclei that create mistier conditions in and downwind of urban areas but reduce rainfall elsewhere.

Following international recognition of the severity of urban atmospheric pollution, there has been increasing effort to reduce concentrations in urban regions throughout the world, with emission regulations applied in many areas, often alongside monitoring networks. As a result, many developed urban areas have pollutant levels generally below national regulatory standards (e.g. Baldasano et al., 2003).

3.8.2 Water pollution

Urban waterways can be heavily polluted, usually from excess of nutrients, sewage or heavy metals (Paul and Meyer, 2001; Hatt et al., 2004). Nutrients and heavy metals usually arrive via surface runoff, either directly from the urban surface or via drainage infrastructure, though some pollutants are deposited from the atmosphere. Many large urban regions also have combined sewer overflows (CSOs), which are sewerage systems that also convey water that enters as runoff after rainfall events. These have a safety design built into them that allows water to be discharged directly into waterways rather than overflowing into the streets when flow capacity is exceeded. Such overflows can have detrimental effects on urban waterways, as sewage entering the rivers greatly increases biochemical oxygen demand (BOD), due to the large numbers of oxygen-demanding microorganisms decomposing the organic material. High BOD lowers available oxygen for other species and can lead to mass mortality of fish and invertebrates for some distance downstream of a CSO discharge point, until sufficiently dispersed. Unfortunately, these types of events can happen quite frequently. As an example, along the River Thames through central London, CSO-related pollution events occur on average once a week during periods of moderate rainfall and have resulted in around 39 million tonnes of untreated sewage being released directly into the Thames annually (Thames Water, 2012). These events have spurred the construction of the Thames Tunnel, which aims to reduce sewage spills by improving the current Victorian CSO system.

Nutrient concentrations, mainly nitrogen and phosphorus, associated with urban runoff lead to cultural eutrophication of receiving waters. The process of cultural eutrophication involves increased primary production and associated algal blooms. In many cases algal blooms will be dominated by cyanobacteria capable of producing a range of toxins (Codd, 1995). These toxins can have multiple and varied effects on human health and are an important consideration for water treatment works that rely on surface water supply. Cultural eutrophication also results in dissolved oxygen depression and hypoxia due to algal respiration and decomposition of dead cells. The ecological consequences of these events are similar to effects from CSO discharges discussed above. In riverine and estuarine systems nutrient export and chronic eutrophication can affect downstream coastal waters and lead to 'dead zones' or vast expanses of hypoxia (Diaz and Rosenberg, 2008).

Within coastal urban regions, water pollution delivered via rivers and estuaries can have a notable impact on coastal and marine ecosystems. The key impacts stem from excess nutrients (see above) to direct ingestion of plastics or other pollutants, causing mortality of invertebrates, fish and birds. Effects can ultimately be far-reaching; for example, the Great Pacific Garbage Patch is composed of materials primarily of terrestrial origin, many of which will float down rivers that ultimately exit at large coastal towns and cities.

3.8.3 Soil pollution

Many of the pollutants found in urban atmospheres and surface waters are also found in soils, where they may be retained for some time depending on the particular soil conditions and the pollutant under consideration. These may be relatively diffusely sourced, e.g. from vehicular traffic or industrial emissions that are deposited onto soil surfaces, but also from direct point sources such as manufacturing sites. Soils on brownfield sites, for example, are noted for their (often) high levels of industrial pollutants, particularly heavy metals such as nickel, copper, lead and cadmium. Development of such sites often requires extensive soil treatments to reduce pollution levels. Heavy metals are particularly common in urban soils, from both industry and vehicles (see Pickett *et al.*, 2011). Many soils are intensively managed, and this may involve the addition of substantial amounts of fertilisers and pesticides to the soil (e.g. Robbins and Berkenholtz, 2003). Santorufo *et al.* (2012) have demonstrated that soil pollutants (among other characteristics) can affect richness and abundance of different soil invertebrates, with taxa having differential responses to pollution levels.

3.8.4 Light pollution

Urban areas are almost universally well-lit. The NASA image of the Earth at night (Figure 3.2), as entrancing as it may be, accurately maps some of the most urbanised parts of the world and demonstrates both how much energy is being used and how much light emitted in urban areas. This can result not just from direct illumination from artificial lighting (e.g. streetlamps or shop lighting) but also from reflection from artificial surfaces such as pavements and buildings. Light pollution does not just reduce the visibility of stars at night, but can also alter animal behaviour patterns, for example birds displaying nocturnal activity often as a result of confusion, and changes to phenology (Longcore and Rich, 2004). Davies *et al.* (2012) demonstrated that artificial lighting may even influence invertebrate assemblages found in lit areas, with predatory and scavenging species (e.g. ground beetles, ants and woodlice) in particular being more abundant around street lighting.

3.8.5 Noise pollution

Everyone knows urban environments can be noisy, but both the level and type of noise in urban areas can have both societal and environmental impacts. 'Noise' can be classified as any undesired sound, but is more technically a persistent sound with an

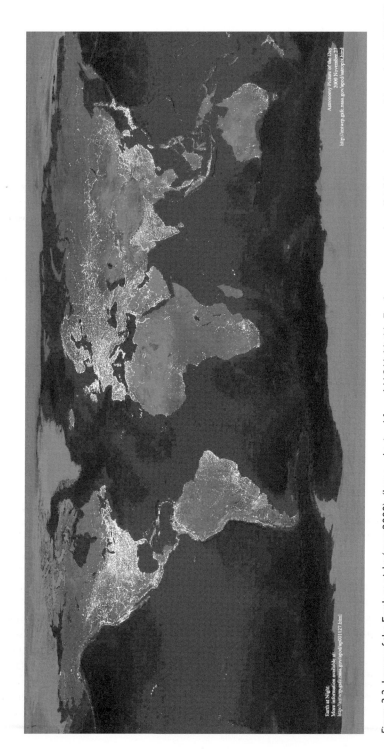

Earth at Night.
More information available at:
http://antwrp.gsfc.nasa.gov/apod/ap011127.html

Astronomy Picture of the Day
2000 November 27
http://antwrp.gsfc.nasa.gov/apod/astropix.html

Figure 3.2 Image of the Earth at night (year 2000). (Image obtained from NASA's Visible Earth programme: http://visibleearth.nasa.gov/view.php?id=55167.)

irregular wave pattern, particularly if it occurs outside the range of more natural ambient sound. Urban noise mainly comes from traffic, aircraft, construction and people, and is generally higher around roads, construction sites and commercial districts. Societal impacts include increased stress, chronic hearing damage, and an overall reduction in quality of life. Environmental impacts include changes to animal behaviour, for example the need for avifanua to alter their birdsong to account for the volume and duration of urban noise (Wood and Yezerinac, 2006; Chapter 6).

3.9 Biodiversity

Patterns of biodiversity within the urban ecosystem can be complex and are driven by many factors. Certainly biodiversity can be compromised in urban regions (e.g. Araújo, 2003), but in some cases, and for certain groups of organisms, diversity can be increased. Biodiversity impacts within the urban region can occur at several spatial scales. Looking first of all at the numbers and types of species found within the urban region, it is clear that urbanisation has a direct effect on habitat provision due to the replacement of surfaces with more impervious land cover that has generally reduced capacity to act as habitat (Arnold and Gibbons, 1996; Rosenzweig, 2003). This, combined with high levels of habitat fragmentation that create smaller and more isolated patches, leads to a reduction in both species richness and populations of many taxa following fundamental species-area relationships (Boxes 2.4 and 2.5), though this can be highly variable (Crooks *et al.*, 2004; Pauchard *et al.*, 2006; McKinney, 2008).

Urban-associated changes initially lead to a decline in more specialist species, particularly those using relatively scarce habitat types. Correspondingly, a general increase in more generalist or opportunistic species tends to occur. Indeed, urban regions can display significant concentrations of species, leading to unexpectedly high levels of biodiversity. Urban biodiversity may be counter-intuitively associated with human population size. Klotz (1990) found a stronger correlation between plant species richness and log total population than with urban area, and Sukopp (1998) found over 1000 species in urban regions with 100,000–200,000 inhabitants. In some cases, urban diversity may be higher than in the surrounding ecosystems, particularly if they are agricultural (Wania *et al.*, 2005). These relationships often hold true at broad scales, though Pautasso (2007) has demonstrated that at fine spatial scales a reduction in diversity is sometimes found.

McKinney (2008) reviewed 105 studies of biodiversity in urban environments that recorded changes in species richness across a gradient of urbanisation. Although particularly dense urban areas displayed a consistent reduction in total species richness, particular groups of organisms responded differentially to more moderate levels of urbanisation (Figure 3.3).

Plant species in particular displayed an increase in species richness in many of the studies when comparisons were made between low (<20 per cent impermeable surface area, ISA) to moderate (20–50 per cent ISA) urban cover (Figure 3.3). These findings may be explained by several factors. Humans are the main vectors for plant introduction into urban regions (intentional or not) and consequently higher human population densities are likely to increase the frequency, intensity and duration of plant introductions ('introduction effort'; see Section 6.3.1) and consequently establishment (Pauchard *et al.*, 2006; Chapter 6). Older urban regions also have

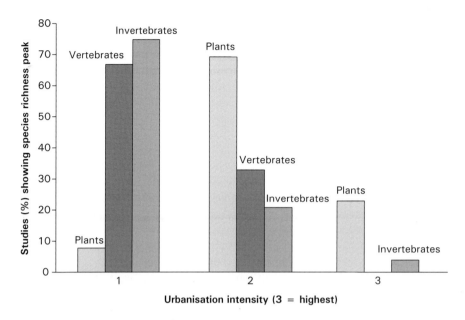

Figure 3.3 Percentage of published studies showing peaks in species richness of plants, vertebrates and invertebrates at different levels of urbanisation: I = <20 per cent impermeable surface area; 2 = 20–50 per cent ISA; and 3 = >50 per cent ISA. (Reproduced from McKinney, 2008, with permission.)

higher plant diversity, for similar reasons – there has simply been more time for intentional or unintentional plant introductions to result in establishment (Kowarik, 1990; Sukopp, 1998). Indeed, non-natives are common in urban areas, and this is discussed in more detail in Chapter 6.

High human population density correlates to urban development and fragmentation, which essentially creates a mosaic of many different habitats with small footprints within urban areas. Forman (1996) demonstrates that a collection of small patches can still contain a reasonable proportion of species. This can allow for urban areas composed of a complex mosaic of small patches to be quite diverse. These patterns can also relate spatially to the disturbance heterogeneity model (see Section 2.5.2; Porter *et al.*, 2001), which explains increased diversity due to a dense mosaic of adjacent habitats that allows for rapid turnover of species and many species living within a spatially limited area.

For non-plant taxa, McKinney (2008) found that the majority of studies (e.g. 82 per cent for vertebrates and 64 per cent for invertebrates) indicated a reduction in species richness at transitions between low to moderate urban cover, with only a small minority showing an increase. Such decreases with urbanisation are probably due to:

1   a reduction in habitat area, which may be more significant for animals (particularly vertebrates), which have larger body sizes and therefore require larger areas following the principles of allometry and habitat use, as they need sufficient habitat to complete their life cycles;

2   a decline in vegetation cover and quality (including structural complexity) in many
    urban areas that compromises habitat resources, often with an increase in non-
    native plant species (see above) that may be unsuitable for many native animal
    species;
3   active management against the species, e.g. pest control;
4   competitive displacement by more dominant generalist or synurbic species (see
    Chapter 6).

Birds in particular are well-documented to display reduced diversity in urban areas
(see Marzluff and Ewing, 2001; Chace and Walsh, 2006; Pauchard *et al.*, 2006), probably because avifauna (particularly, e.g. insect-feeders) have been shown to be particularly sensitive to habitat fragmentation (Forman, 1996).

The pressures of urban environments will alter species population levels and persistence, resulting in changes in the plant and animal assemblages found (see Chapter 6).
Urban environments are mainly dominated by generalists that have flexible responses
to novel forms of resource availability (Francis and Chadwick, 2012), though some
species are favoured by the creation of artificial 'habitat analogues', such as rock/cliff
species within the built environment (see Lundholm and Richardson, 2010; Francis and
Hoggart, 2012). Urban organisms may also display altered environmental responses
(e.g. Francis and Chadwick, 2012), which has sometimes been linked to processes of
natural selection and microevolution (e.g. Badyaev *et al.*, 2008). This is discussed more
in Chapter 6. The spatially heterogeneous and rapidly changing urban environment
inevitably results in changes in interactions between organisms, and between organisms
and their environment. Changes in resource availability and inflated productivity, along
with reduced predator–prey interactions, reduction of seasonality and sometimes simplified assemblages will affect urban food webs and trophic dynamics (Shochat *et al.*,
2006), for example.

3.9.1 Socio-economics and urban biodiversity

Economic status has often been found to correlate with both environmental quality
and access to environments that people find pleasant. However, it is not as simple as
richer people living in more biodiverse areas, particularly when trends of gentrification and redevelopment are considered, which over time may obscure any relationships between economic status of previous residents and biodiversity or green space
(Warren *et al.*, 2010). Having perhaps inadvertently shaped patterns of biodiversity
and environmental quality, the affluent may then blur them again with patterns of
redevelopment.

Economic status certainly correlates to factors that are likely to influence the urban
environment, such as housing density, garden presence and size, proximity to public
parks or other green space, as well as tendency to engage in horticulture, plant non-
native species and so on (e.g. Troy *et al.*, 2007). This has been shown to be important
in determining patterns of biodiversity in cities, sometimes interrupting more traditional patterns assumed to reflect, for example, population and built structure density,
distance from urban centre (rural-urban gradient), and time since urbanisation (Hope
et al., 2003; Kinzig *et al.*, 2005; Pauchard *et al.*, 2006).

In particular, there is a strong spatial relationship between economic status and abundance or proximity to green space. Strohbach *et al.* (2009) found that high-income residential areas were located near parks, woodland or river ecosystems in Leipzig (Germany), and consequently had generally higher levels of bird species richness. However, such trends are not always straightforward, with, for example, observations from Merseyside (UK), that more affluent areas lost greater amounts of green space during 1975–2000 than less affluent areas, mainly due to gardens being paved over or built upon; though less affluent areas also lost green space during this period due to the redevelopment of brownfield sites (Pauleit *et al.*, 2005).

Private green space in particular is linked to economic status, particularly gardens. Garden size and abundance for example are closely related to economic status and associated indicators such as house size and density, influencing the diversity they may support. Tratalos *et al.* (2007b) found that social status was linked to increased tree cover in five urban regions of the UK, mainly due to associations with gardens or other green space. Hope *et al.* (2003) found a positive correlation with household income and plant diversity in Phoenix, Arizona, with diversity values above the median income for the area ($50,750 in 2000) being twice as high as those below, on average. This may have been due to intentional planting and cultivation of a wider range of species, along with proximity to green space, and/or the tendency of higher income residences to be positioned at higher elevations in this semi-arid region, which are naturally more biodiverse. In this case, more biodiverse areas may have been 'acquired' rather than 'constructed' by the more affluent residents, though a combination is probably the case.

Increased income and private ownership certainly means increased choice and selection of species for private green space (which Hope *et al.* (2003) term the 'luxury effect'), and this can create differences in assemblages between areas differing in socio-economic status (e.g. Sudha and Ravindranath, 2000). Indeed, socio-economic status is also related to the desire to enhance private green space or participate in other ecological improvements. Although some forms of habitat enhancement do not require substantial investments of resources at an individual level (e.g. wildlife gardening), some do (e.g. living roofs/walls), and economic status also correlates to other factors likely to influence enhancement potential, such as housing density, garden availability and size, proximity to public parks, patterns of home ownership and other social factors such as level of education and crime (e.g. Troy *et al.*, 2007). As Kinzig *et al.* (2005) note, groups with higher economic status can devote more resources to 'creating their ecological ideal' (p. 2), if they are motivated to.

Such patterns are usually described for vegetation and plant diversity (e.g. Hope *et al.*, 2003). However, socio-economic trends affect animals too, either directly or indirectly via differences in habitat provision. For example, Smallbone *et al.* (2011) found that the diversity of frog species in urban regions of Victoria and New South Wales was determined by urban intensity, cover of native vegetation and isolation of wetland habitat; of these variables, vegetation cover was significantly related to economic status (for the reasons noted above, e.g. greater size and abundance of private gardens), meaning that such areas maintained higher diversity of frogs in general – despite lower status areas being located in lower elevations that have a greater abundance of natural wetlands.

Socio-economic factors are therefore a further important driver of species distributions and diversity in urban regions, though it should also be remembered that urban ecological assemblages are spatially complex even in areas that may be similar in their physical and social organisation, partly due to spatial heterogeneity present but also reflecting a localised history of species introductions and management by residents, as well as former land use (Hope *et al.*, 2003; Lubbe *et al.*, 2010). As such, these remain generalisations only, but are nevertheless of use when trying to understand patterns of diversity and for spatial planning.

3.9.2 Key trends in urban biodiversity

Further discussion of urban species and biodiversity is found in Chapter 6, but key trends found across the urban ecosystem include:

1   an increase in more generalist species and a reduction in specialists (see Section 6.2);
2   a general decrease in biodiversity at all levels, including globally due to biotic homogenisation, but occasionally localised and regional increases in diversity of some species groups (and sometimes overall biodiversity) in comparison to pre-urban or non-urban ecosystems;
3   an extinction time-lag, particularly in recently urbanised areas, where biota are still adjusting to fragmented habitats and changes to resources, biotic interactions and so on – some urban biodiversity is therefore remnant from an earlier land-scape structure, and will go locally extinct in time (Hahs *et al.*, 2009);
4   an increase in numbers and proportions of non-native species (see Section 6.3);
5   a decrease in overall populations of many species but sometimes an increase in population density;
6   disrupted and atypical seral processes, which are made complex by differential species availability and changes in stress and disturbance (see Section 6.5);
7   local environmental factors are often more important in driving diversity and assemblage composition than landscape scale factors, though landscape position may sometimes be important, especially for certain species groups – this is particularly the case for source populations within the landscape, for example (Westermann *et al.*, 2011).

This is only an overall summary of some of the ways in which biodiversity and ecological communities may be affected in urban environments. The next three chapters discuss the various ecosystems found within urban environments in more detail (Chapters 4 and 5) and the types of species found (Chapter 6).

3.10 Chapter summary

• Urban ecosystems are complex adaptive systems characterised by: (1) multiple components that interact across a range of scales; (2) non-linear patterns and processes; (3) existence in a state apart from equilibrium that requires a flow of energy for maintenance; (4) self-organised properties; and (5) the capacity of both components and the system to 'learn' or change in response to environmental factors.
• Ecosystem services are important within urban ecosystems. These are the processes and goods that ecosystems supply that benefit humans, and include

provisioning, regulating, cultural and supporting services. It is also important to consider disservices, which are harmful or detrimental processes.

- Urban biogeochemical cycles are altered in urban regions, reflecting land-use changes caused by urbanisation but also urban emissions and rates of deposition. Cycles of nitrogen, phosphorus and carbon are the most scrutinised.
- Urban climates are characterised by increases in temperature relative to non-urban areas (the urban heat island effect), and changes in precipitation patterns. In arid locations, urban areas may be cooler than the surrounding non-urban land. These climatic effects are linked to changes in phenology and animal behaviour.
- Hydrological cycles are usually heavily disrupted in urban areas, with impermeable surfaces causing a decrease in infiltration and evapotranspiration, and an increase in surface run-off. This creates problems with both urban flooding and water deficit, with concomitant ecological impacts.
- Urban soils are generally artificial and highly disturbed, with the exception of some green-space soils. Despite this, they are important for certain species groups and urban ecosystem services.
- Urban areas often have high levels of pollution, including atmospheric, aquatic, soil, light and noise. These all have impacts on urban species as well as humans. More developed regions generally have more pollution regulation and controlled environments than those undergoing rapid urbanisation.
- Patterns of biodiversity are complicated in urban regions, with losses of many species and groups overall, but increases in others due to spontaneous colonisation and human introduction. Almost all species decline with severe urbanisation (>50 per cent impervious cover), but some species and groups (e.g. plant taxa) increase with moderate urbanisation (20–50 per cent impervious cover).
- Socio-economic factors also influence patterns of biodiversity, including green-space provision and the ability to select species, e.g. for planting.

3.11 Discussion questions

1   Consider the different flows into and out of the urban ecosystem. Which are likely to be the most important for global environmental impacts, and why? Which might be the easiest and hardest to regulate?

2   Which ecosystem services do you consider most important within an urban context?

3   How polluted do you consider your urban region to be? Consider the different forms of pollution found in your region. Are any particularly significant?

4   How at risk of flooding is your urban region? What factors are likely to influence this?

5   Why is it important to understand patterns of biodiversity in urban regions, and what drives these patterns? Are you surprised by increases in the diversity of some organisms in urban regions? Why do you think the majority of studies in urban regions have focused on plants and larger animals such as birds and mammals?

Chapter 4

Ecosystems within urban regions

Green space

4.1 Introduction

'Green space' usually refers to any open vegetated area within an urban environment, and is often synonymous with 'open space', though the latter does not explicitly indicate the presence of vegetation (Sadler *et al.*, 2010). Here, green space refers to sections of the urban environment that retain some characteristics typical of natural or semi-natural ecosystems, for example vegetation cover, or which are regenerating spontaneously (such as brownfield sites), as opposed to those areas that are primarily associated with the built environment (Chapter 5). Consequently, this chapter covers parks and recreational spaces, gardens, lawns, allotments, brownfield and wasteland sites, and areas of remnant woodland. Urban rivers and lakes are also covered here, as although they are often heavily impacted they are not truly artificial, and often contain vegetation and other characteristics often associated with green space, such as important habitat provision and desirability of residential proximity.

Urban green space can be abundant. Estimates of surface cover of green space range from 2 per cent to 46 per cent of surface cover within Europe (Fuller and Gaston, 2009). In the UK, green space on average takes up 14 per cent of urban areas (Sadler *et al.*, 2010). Green space is often quantified as m^2 per capita, partly in recognition of its amenity value and also to allow comparison between and within urban regions (e.g. Jim and Chen, 2008; McConnachie and Shackleton, 2010). Jim and Chen (2008), for example, note that Taipei (Taiwan) maintained approximately 4.95 m^2 of green space per capita in 2004, compared to 3 m^2 per capita for Hong Kong and 7 m^2 per capita for Singapore (see Box 4.1). Fuller and Gaston (2009) have determined that in many cases the coverage of green space is linked to urban complex size rather than population, so that smaller complexes tend to have lower per capita green space provision. Of course, it is not just the abundance of green spaces that is important, but also their ecological quality. As noted in Chapter 2, increased urbanisation is linked to increased green space fragmentation, which reduces ecological quality because of reduced areal coverage and increased patch isolation (see Chapter 2).

Individual types of urban green space are now considered in more detail.

4.2 Public parks and recreational spaces

Parks and other public recreational spaces (e.g. school fields, sports pitches, university campuses, playgrounds; Figure 4.1) are among the largest and most contiguous

Box 4.1 Singapore: a garden city

Singapore is widely perceived as a 'garden city' with abundant green space – a feat which is all the more remarkable given the urbanisation pressures faced by the self-contained island state (Tan, 2006). The original rainforest ecosystem present before urbanisation has largely been lost, despite some early attempts to conserve it. Singapore originally contained many areas of rainforest and mangrove forest established as nature reserves during British rule in the late nineteenth century, though legislations protecting the reserves were revoked and reinstated several times to allow for urban development, even after colonial rule ended in 1963. Currently only three rainforest reserves remain, though this still covers a sizeable total land area of 2138 ha (Ooi, 2011). But Singapore has been highly proactive in encouraging the creation of green space. Urban development has mainly been in the vertical dimension, with multistorey buildings constructed to ensure that current urban space is used most efficiently.

Much of the urban greening of Singapore has occurred in the last 50 years, from initial attempts to plant trees alongside urban infrastructure to the full intention to create a 'garden city', with the formal creation of a Garden City Action Committee in 1973, which still exists (Tan, 2006). Initiatives have resulted in an abundance of parks and street trees, as well as the creation of living roofs or 'roof gardens' (Yuen and Hien, 2005). A benchmark aim of 0.8 ha of green space per 1000 citizens has been established by the Garden City Action Committee, with Yuen and Hien (2005) recording an availability of 0.6 ha per 1000 citizens in 2005. Singapore certainly compares favourably to many other areas, with 7 m^2 of green space per capita in 2004 (Jim and Chen, 2008).

Singapore has also been forward-thinking with respect to spatial planning and networks of green space. Tan (2006) presented a green network plan for Singapore (see Figure 8.3), noting that clear guidelines exist in terms of connecting greenway corridors (e.g. size), and that around 360 km of greenways are planned for the next two to three decades, mainly connecting key park areas. Such a network should improve the living environment for residents as well as increasing biodiversity. Singapore may be an excellent model for marrying urban greening with compact urban development, and a refreshing alternative to ongoing urban expansion elsewhere. It also highlights what can be achieved with appropriate political and societal support for green space planning.

vegetated areas within urban regions. Total land cover of parks/recreational spaces varies notably among regions. Jim and Chen (2008) indicate that urban parks account for 1.8 per cent of Hong Kong and 2.5 per cent of Singapore, while Keefe and Giuliano (2004) note that parks cover around 13 per cent of New York City. Variation in extent will also depend on the way parks and recreational spaces are classified; some studies include 'reserves' or fragments of relatively pristine habitats, while others only consider managed recreational spaces as 'parks' or 'parkland'.

In some cases these spaces are intensively managed but may contain remnants of pre-urban ecosystems (e.g. woodland patches – Clarkson *et al.*, 2007; Jim and Chen,

Figure 4.1 An urban park in the United Kingdom, characterised by different vegetation types, including large standing trees, lawn or grassland areas and planted strips and borders. Tracks and recreational facilities make parks an important facility for urban residents.

2008) that support biodiversity otherwise lost (Natuhara and Imai, 1999). Parkland usually contains a range of vegetation types, from semi-natural grassland to lawn, to woodland and individually planted trees, as well as water features and sometimes elements of the built environment. Many urban ecosystems are too small to contain sections that are not influenced by an edge effect (Forman, 1996; Hostetler and Drake, 2009; see Box 2.3). However, large parks that contain fragments of natural habitat are among those most likely to maintain 'interior' species (Natuhara and Imai, 1999; Drinnan, 2005). Large parks are also likely to have higher levels of habitat heterogeneity, which are important for allowing the coexistence of a wide range of species (e.g. Porter *et al.*, 2001; LaPaix and Freedman, 2010).

At the landscape scale, patch isolation and influence of the urban matrix have also been shown to be important for park diversity. For example, Oliver *et al.* (2011) found that richness of breeding and wintering birds in urban parks of the greater St. Louis area, Midwest, USA, was determined by both park area and the level of urbanisation in a 5 km radius. These relationships showed how biodiversity in urban areas can be affected by the relative isolation of parks. Park location within the urban region may also influence the biodiversity and species found. MacGregor-Fors and Ortega-Alvarez (2011) found that the main predictor of both species richness and abundance of avifauna in Mexico City was distance from the edge of the urban complex. These factors do not apply to all taxa of course; for example Clarke *et al.* (2008) found no relationship between park characteristics and ant diversity in urban parks of San Francisco.

Few studies have evaluated the biodiversity of multiple species groups or assemblages in parks and recreational spaces, instead focusing on usually well-documented groups such as birds (e.g. Fernández-Juricic and Jokimaki, 2001; MacGregor-Fors and Ortega-Alvarez, 2011; Oliver *et al.*, 2011), mammals (Sorace, 2001; Mahan and O'Connell, 2005), plants (Loeb, 2006; LaPaix and Freedman, 2010), and popular invertebrates (Giuliano and Accamando, 2004; Clarke *et al.*, 2008; Clergeau *et al.*, 2011). Much of the diversity found in urban parks is context-dependent, on park size, position, level of urbanisation, structure, complexity, biogeographical region and so on. Nevertheless, general trends are that urban parks contain higher diversity than the surrounding urban complex for most, though not all, species groups, regardless of biogeographical location of the urban region. This can be illustrated in plant communities by the fact that most parks support a mix of spontaneous colonisers from the wider urban environment, remnant species that remain from pre-urban ecosystems and species planted for horticulture (Hermy, 2010). Similarly, parks within the urban complex of Flanders, Belgium, contained between 30–60 per cent of species from different groups found within the wider region (Cornelis and Hermy, 2004). Their value for urban biodiversity as well as ecosystem services is apparent, and this is one reason why much urban conservation and ecological planning has focused on these ecosystems, either individually or as networks of green space (Chapters 8 and 9).

Parks are generally public recreational spaces. Though less public, golf courses also represent relatively large recreational spaces in urban regions, and provide habitat for a range of species (e.g. Yasuda and Koike, 2006; Colding and Folke, 2009). Saarikivi *et al.* (2010) recorded 72 species of carabid beetle in urban golf courses of Helsinki, Finland, and found distinct assemblages on individual courses. Some golf courses actually maintained remnant species present from before development, but specialist species were rare. Colding *et al.* (2009) note that golf courses often contain freshwater or wetland habitat, which can be rare in urban environments. They found that aquatic invertebrate and amphibian diversity on golf courses were comparable or superior to other urban parkland. However, other animals also take advantage of golf course habitats. Cornell *et al.* (2011) found that Eastern Bluebirds (*Siaila sialis*) in Virginia experienced more reproductive success on urban golf courses compared to other urban or agricultural reference sites. Perhaps the main ecological benefits of golf courses result from their size (as they tend to cover a lot of land) and their habitat heterogeneity. By design, golf courses contain a range of vegetation types (from lawn/turf to woodland) and water features which have been shown to support a range of species less tolerant to urbanisation (Hodgkison *et al.*, 2007). Some fairways, for example, provide tall grass and scrub habitat that is less well-managed than most urban parks, but is not comparable to brownfield vegetation, and may therefore represent a habitat or a seral stage that is not easily found elsewhere. Perhaps the most important ecological aspect of golf courses is the range of ecosystem services they can support. These include carbon sequestration and greenhouse gas fixation, regulating urban runoff and associated flooding, and of course recreational and human health benefits.

4.3 Gardens

Urban domestic (private) gardens can cover a substantial proportion of the urban region. Studies have found gardens covering large urban land areas, for example approximately 16 per cent for Stockholm, Sweden (Colding *et al.*, 2007); 19 per cent for Santiago, Chile (Reyes-Paecke and Meza, 2011); 23 per cent for Sheffield, UK

(Gaston *et al.*, 2005a); and 36 per cent of the urban area of Dunedin, New Zealand (Mathieu *et al.*, 2007). Many gardens are individually small in area but very abundant, while large urban gardens are relatively rare (Gaston *et al.*, 2005a). Gardens have recently attracted quite a lot of attention both as ecosystems with relatively high biodiversity for urban regions (Thompson *et al.*, 2003; Goddard *et al.*, 2010) and areas that can be promoted for biodiversity by the improvement of local habitat characteristics (Gaston *et al.*, 2005a, 2005b). Davies *et al.* (2009) estimated that 87 per cent of homes in the UK have a garden, in total containing around 2.5–3.5 million ponds and 28.7 million trees. The ecological value of gardens is substantial.

Gardens come in a wide variety of forms – from simple grass patches to highly diverse collections of native and ornamental plants. The most detailed work on urban gardens has come from the 'Biodiversity in Urban Gardens in Sheffield' (UK) research programme. This project examined 61 gardens of varying size throughout Sheffield to examine their role in urban biodiversity (Thompson *et al.*, 2003; Smith *et al.*, 2006a). From this sample, 1166 species were recorded, of which 30 per cent were native and 70 per cent were alien. This did not cover the entire garden plant diversity within the urban region, however, as new species were still being found even at the end of the survey (Smith *et al.*, 2006a). Native species occurred more frequently throughout the gardens – there were more alien species overall, but individual aliens occurred less frequently. The species richness of plants in the sampled gardens closely matched the uncultivated species richness of the entire examined urban region, which clearly demonstrates the diversity that can be found in gardens (Smith *et al.*, 2006a). Mean garden species richness was 112 species, though variation between gardens was high (standard deviation of the sample being ±50 species), reflecting differences in garden size and range of habitat types. The native/alien balance was the same regardless of garden size, though garden area was significantly positively related to species richness, primarily for natives. The species–area relationship for gardens is not simple, however. Plant diversity of gardens has been linked to cumulative garden area at the landscape scale (Thompson *et al.*, 2003), but this is not always the case for individual gardens (see also discussion for lawns (Section 4.4), which do not always show a positive species–area relationship, and may therefore complicate general trends for gardens). Further, Goddard *et al.* (2010) note that individual gardens alone are probably too small to maintain viable populations of most species.

The ambiguity of the species–area relationship for gardens is further reinforced by Smith *et al.*'s (2006b) observation that invertebrate diversity in gardens is linked to the presence of certain habitat types rather than garden area. However, as would be expected, larger gardens did tend to have greater habitat heterogeneity and higher biodiversity. Invertebrate species richness was generally higher in gardens with trees, and which had abundant 'green space' (including other gardens) in the surrounding 1 ha (Smith *et al.*, 2006b). Specific taxa responded positively to certain habitat types, for example the presence of compost (Coleoptera), ponds (Tipulidae, Pediciidae, Limoniidae and Ptychopteridae) and low canopy vegetation (Apoidea). This highlights the value of maintaining a range of plant functional groups and growth forms in urban gardens, which help to create habitat complexity (Smith *et al.*, 2006a). Of course, richness is also determined by broader environmental factors (e.g. altitude), garden shape (e.g. perimeter/area ratio) and management (e.g. frequency of weeding, tendency of owners to feed wildlife) (Smith *et al.*, 2006b).

Garden characteristics are closely allied to those of the surrounding built environment (e.g., house plot size and housing density). Smith *et al.* (2005) observed that garden size was negatively correlated to the number of houses, total area of buildings and total area of roads in the surrounding 1 ha. Essentially, areas with less densely spaced houses have more total garden area and larger gardens per house. In urban regions, housing density is often correlated to income levels and this typically results in wealthier citizens having greater garden 'resource' to use as they see fit. House construction style was also important; terraced houses contained smaller gardens than semi-detached, which in turn had less garden space than detached houses. This trend partly reflects age of housing, with terraced houses usually being of greater age in Sheffield, but also income levels, as the terraced–detached gradient often relates to increasing house prices and desirability.

Garden shape may also be important in the way they act as habitat. They are often regular in shape, which lends them to an increasing proportion of edge effect, particularly for small gardens. This may be further reinforced by the tendency of owners to maintain planted borders around edges, and also due to the situation of many 'boundary' features at garden edges, such as hedges, fences or walls. All of these edges are potential 'filters' for the movement and colonisation of species within garden patches (e.g. Smith *et al.*, 2005).

Despite their abundance and desirability, gardens are perhaps amongst the most threatened green spaces. As pressure increases for living space in urban regions (for example within the coalescence phase of the diffusion–coalescence cycle described in Section 2.4), gardens are often perceived as land that could be better used to build housing extensions and thereby turned to an economic advantage. In some cases, increasing private car usage, infrastructural pressures, conversion of houses to flats with increased household densities and various other factors have led to difficulties in residents obtaining sufficient public parking space, with a common response being to pave over gardens to create or expand a private driveway (e.g. Peffy and Nawaz, 2008). In response to this, and increased recognition of their ecological value, gardens are now featuring in urban conservation initiatives in some locations (see Chapter 7).

4.4 Lawns

Many recreational spaces, particularly gardens and parks, can have a substantial proportion of lawn area (Figure 4.2). Lawns are usually composed of 2–3 dominant grass species, mainly those included in commercial turfgrass seed mixtures, along with some low-frequency herbs (Stewart *et al.*, 2009). Robbins and Birkenholtz (2003) examined the extent of the 'lawnscape' in Franklin County, Ohio, and found that it accounted for 23 per cent of the land cover. They termed this process of replacing semi-natural vegetation with turf across large areas the 'turfgrass revolution'. Although lawns may provide important ecosystem services such as carbon sequestration, they may also emit N_2O and CH_4, especially when irrigated and fertilised (Livesley *et al.*, 2010). Lawn biodiversity is generally thought to be limited. However, Stewart *et al.* (2009) document 127 species sampled across 327 lawns sampled in Christchurch, New Zealand, though the majority (80 species) were found in less than 2 per cent of the lawns, indicating the overall dominance of relatively few species. Similar observations were made by Thompson *et al.* (2004), who found 159 plant

Figure 4.2 An urban lawn at a residential plot in the United Kingdom, showing the well-managed monoculture typical of most lawns, along with strict hedge borders and occasional standing trees. (Photograph by Rob Francis.)

species across 52 garden lawns in Sheffield, UK, 60 of which were only recorded once (with a mean species richness of 24). Larger lawns were also more likely to contain rarer species. In this sense, lawns may be more similar to semi-natural grassland eco-systems than other garden areas, and may consequently function to some extent as an 'analogue habitat' within the urban environment (see Lundholm and Richardson, 2010; Section 7.7).

Lawns have been shown to display contradictory relationships between species richness and area; Thompson *et al.* (2004) found that larger lawns had higher species richness, while Stewart *et al.* (2009) found that the larger lawns of public parks were less diverse than smaller private lawns. Contradictions may be found in other areas too, probably reflecting planting and management practices; Thompson *et al.* (2004) found a minimum of 83 per cent native plant species composition in the urban lawns of Sheffield, while Stewart *et al.* (2009) found an average of only 13 per cent native composition. Somewhat counter-intuitively, increased management (e.g. by mowing) may lead to increased species richness, and it has been observed that differential man-agement regimes also create distinct 'lawn' assemblages (Stewart *et al.*, 2009). Position within the urban landscape, in relation to other green space and socio-economic patterns, is also likely to be important in influencing lawn diversity and assemblages, as for gardens, though there has been limited investigation of this.

4.5 Allotments

Allotments are areas of public land that are leased by private individuals for fixed per-iods of time for cultivation. These plots used to grow food are common in some urban regions, particularly in Europe, where private space for cultivation is limited. Allotments are found in a range of locations, and even on brownfield sites (Megson *et al.*, 2011). The demand for allotments in Europe probably stems from an agricultural legacy and from the desire for food security. This was magnified during the First and Second World Wars even though the allotment systems have existed for centuries (Flavell, 2003). Although such areas may be important in future food security and sustainability within urban regions (e.g. Leake *et al.*, 2009; Nabulo *et al.*, 2010), their ecological value has yet to be fully investigated. In many ways they may function as gardens, but are somewhere between gardens and brownfield sites in their environmental characteristics.

A few studies have highlighted the ecological and social value of allotments. Several studies have shown how animals benefit. Ahrne *et al.* (2009) found that urban allotments are important for urban bee species, particularly where the abundance of flowering plants was high. Davison *et al.* (2009) noted their value as stepping-stones for the movement of urban badgers. Allotments have also been positively related to house sparrow (*Passer domesticus*) density (Chamberlain *et al.*, 2007), and may provide a useful food source for urban foxes (*Vulpes vulpes*) (Contesse *et al.*, 2004). Social benefits have mainly been related to their communal aspects. Barthel *et al.* (2010) found that allotment gardens in Stockholm, Sweden, were sites where ecological knowledge and practices were shared, refined and imparted across the communities and generations of those using the allotments, as well as providing ecosystem services common to urban green space. In particular the latter may include human health benefits associated with tending the allotments (Leake *et al.*, 2009).

4.6 Brownfields and wastelands

Brownfield or 'wasteland' sites are post-industrial or abandoned areas that for whatever reason have not been redeveloped (Figure 4.3). In some cases this may result from land abandonment or de-urbanisation (Reckien and Martnez-Fernandez, 2011), but often they have high levels of contamination that preclude redevelopment for residential or recreational purposes without significant and expensive remediation (e.g., soil washing, installation of plants or microorganisms that may help to break down or extract pollutants, or complete removal and replacement of contaminated soils). The characteristics of brownfield sites ensure disuse by humans, and consequently they become more available to other organisms. Along with potentially high levels of soil and water pollution, brownfield sites often have low nutrient conditions that select for species tolerant of such environments – many of which are declining in their natural ranges due to our tendency to increase nutrient inputs and reduce stressful conditions in many of our anthropogenic landscapes, such as arable fields or pasture (Chipchase, 1999; Eyre *et al.*, 2003). As such, brownfield sites represent key habitat for supporting some species, and indeed it is not uncommon for rare species or even those that were thought to be extinct to be found in urban brownfield areas (Eyre *et al.*, 2003; Jones, 2006). Individual patches may support a high proportion of the diversity of the urban region; Muratet *et al.* (2007) for example found that 58 per cent of the floral richness of Hauts-de-Seine, France, was expressed in urban brownfield sites. These findings have resulted in a move towards the recognition of the ecological value of such sites and their incorporation into urban biodiversity management and conservation plans

Figure 4.3 A small urban wasteland site in the United Kingdom, in this case with a rubble and coarse soil substrate. Note abundant spontaneous vegetation growth and mixture of herbaceous, shrub and tree species. The invasive species (in the UK) *Buddleja davidii* is abundant on this particular site. (Photograph by Rob Francis.)

(Kattwinkel *et al.*, 2011). In some situations where brownfield sites are redeveloped, brownfield habitat may be retained or re-created in order to support those species that often associate with such sites (see Box 7.1); and indeed, much of the value of urban living roofs is as replicated brownfield habitat (e.g. Bates *et al.*, 2009; Francis and Lorimer, 2011). The extent of brownfield sites makes their value even more important – estimates of areal coverage of brownfields have included 80 ha (6 per cent) for Bari, Italy (Lafortezza *et al.*, 2008) and 10,000+ ha in Scotland, UK (Macadam and Bairner, 2012), while over 2 million ha was estimated in the USA in 2003 (Schenck, 2003–4).

Pollution levels at brownfield sites may affect the development of species assemblages and determine seral trajectories. Gallagher *et al.* (2011) found that the distributions of tree species were positively associated with heavy metal contaminants, while herbaceous species were negatively associated. In both cases, responses were not evident until thresholds of metal contamination were met. As a result, different seral trajectories were dependent upon pollutant loads. However, many sites do not have pollutants at levels that impact plants or other organisms, and so impacts are highly site-specific (e.g. Murray *et al.*, 2000).

Beyond contamination, natural soil conditions and associated nutrient levels are important for determining diversity and assemblage composition in brownfield sites. Soil quality can strongly be influenced by past industrial use, with, for example, Schadek *et al.* (2009) finding substantial differences between sites that did or did not contain brick rubble in the soil in Bremen and Berlin (Germany). Though brownfield soil development can be rapid, with Howard and Olszewska (2011) recording the development of a 16-cm-thick topsoil with 2 per cent organic matter content after only 12 years following site abandonment at sites in Detroit, USA, distinct trends of soil development associated with site age are difficult to establish (Schadek *et al.*, 2009), further reinforcing the importance of initial site conditions. Likewise, although there have been some limited investigations of brownfield soil nutrient conditions (Strauss and Biedermann, 2006), these are likely to be highly variable, and dependent on soil material, organic content, type of previous industrial use, time since abandonment, and other factors. Nutrient levels will clearly have some influence on the species assemblages that establish on brownfield sites, but specific investigations are rare.

A range of factors drive diversity on individual brownfield sites. In a study from Germany, Schadek *et al.* (2009) found that plant diversity was positively associated with vegetation height but negatively linked to vegetation density (both of which increased with time since site abandonment), soil phosphorus content and soil water availability, and concluded that diversity is highest when early to mid-sere species are present. This echoes findings by Kattwinkel *et al.* (2011), who determined that brownfield sites also in Germany supported the most species in their first 15 years from cessation of industry, and recommended a shifting mosaic of land abandonment and redevelopment in urban areas to support biodiversity. In many cases, local factors such as vegetation structure, habitat heterogeneity and soil characteristics, many of which relate directly to site age, have been found to be related to diversity and differences in species assemblages rather than landscape scale factors such as position along an urban gradient or proximity to green space (e.g. Small *et al.*, 2006; Strauss and Biedermann, 2006).

4.7 Remnant and regenerated woodland

Many urban regions contain remnant woodland in small, but not particularly abundant, patches; though estimates of cover are uncommon. Cover is usually highest in the fragmented 'edges' of the urban complex, and usually declines with increasing proximity to urban centres (Jim, 2010). In the 100 largest urban regions of Sweden, Hedblom and Söderström (2008) found an average woodland cover of 20 per cent with a range of 1–40 per cent. However, a large majority (around 95 per cent) was found in exurban areas. Individual patches of woodland may range in size from 1 to 2 ha, found within other forms of green space such as parkland, to extensive areas that cover hundreds or even thousands of hectares. For example, Gundersen *et al.* (2006) note that the average size of urban woodlands in Norway was 1000 ha, though this did include woodland on the edge ('fringe') of the urban complex. Jim (2010) notes that urban woodland is characterised by relatively dense tree cover that is not actively managed, as opposed to the sparse tree cover or regulated distributions found in urban parkland. Remnant and regenerated woodlands are therefore more similar in many aspects to semi-natural woodland that may be present in exurban or non-urban areas, and as a result may have particular ecological and societal values. Urban woodlands are of course not entirely natural – their close proximity to urban areas will have created a legacy of management covering decades to centuries in many cases, and their remnant nature reflects only that they have not been directly converted to urban land cover. This may often occur where the urban landscape is less conducive to building (for example on hill slopes), where the woodland has a particular cultural significance (such as a religious site, or because the woodland provides important communal services or resources), or where the desire to have sections of woodland within the urban area was incorporated into urban planning, particularly in more recently urbanised areas (Konijnendijk, 1999; Jim, 2010).

Remnant woodland is more likely to retain potentially rare species (within an urban context), particularly where the woodland has retained a reasonable size. However, most urban woodlands have been influenced by an edge effect, reducing interior species and increasing urban-associated species (Godefroid and Koedam, 2003; Alvey, 2006; Hedblom and Soderstrom, 2010; Ivanov and Keiper, 2010). Hedblom and Söderström (2010) found higher numbers of tree-nesting species in urban woodland compared to the surrounding land use in 34 Swedish urban regions, indicating that the loss of urban woodland would significantly reduce the regional populations of these species. Natuhara and Imai (1999) estimated that interior woodland avifauna in Osaka Prefecture, Japan, needed patch sizes of at least 20 ha to support populations. Indeed, diversity and assemblage composition may be influenced more by fragmentation and reduction in woodland size, and consequent changes in local habitat availability, rather than specific urban impacts (Niemelä et al., 2002). The positive influence of edge effect has been demonstrated by Ivanov and Keiper (2010), who observed higher ant species richness in urban forest edges compared to interior areas in woodland located in Cleveland, Ohio. However, this study also found that forest-associated species were found evenly throughout woodland interior and edge while generalist species had higher diversities in edge environments. Such trends may mean that although species typical of the pre-urban ecosystem may be lost, overall species diversity may not be reduced as much as expected or may even increase (e.g. McKinney, 2008). This phenomenon is not universally observed. For example, Godefroid and Koedam (2003) found that on average more remnant species were found in woodland edges than in the woodland interior (23 per cent compared to 5 per cent) around Brussels, Belgium, though this may have been due to the edge not being old enough for plant assemblages to have adjusted to the edge environment.

Urban woodlands are also likely to contain non-native species, particularly around the edges (Zipperer et al., 1997; Godefroid and Koedam, 2003; Motard et al., 2011; Nemec et al., 2011), further influencing diversity. Some remnant woodland may therefore have higher species richness or diversity than surrounding woodland within the region, or even contain rare or endangered species (Alvey, 2006). This diversity is promoted further by the structural and habitat heterogeneity associated with woodland patches, which may of course incorporate standing dead wood, grass clearings, aquatic habitat and so on (e.g. Natuhara and Imai, 1999; Godefroid and Koedam, 2003; Jim, 2010).

All woodland areas, particularly small ones, can be heavily influenced by the surrounding urban matrix, and such effects will vary according to species groups under consideration. This can be seen in Croci et al. (2008) who found that less vagile species (e.g. beetles) within urban woodlands were more severely affected by their context of urbanisation than more vagile species (e.g. birds). However, even species that can easily disperse are influenced by variations in structural and habitat heterogeneity in woodland ecosystems.

Not all urban woodland is formed by remnant, undisturbed patches. Jim (2011) notes that dense unmanaged woodland in urban areas is also found where succession has occurred on brownfield or wasteland sites. Bornkamm (2007) examined spontaneous woody vegetation regeneration on heavily modified urban soils, finding that trees began to dominate between 14–25 years after colonisation. This suggests that dense regenerated woodland may occur over a few decades given the right conditions (e.g., sufficient colonisation and lack of disturbance or management). Therefore the potential for this process to create woodland patches at different seral stages within many urban regions is evident.

Regeneration of urban woodlands can also be highly managed. Occasionally, large areas of more-or-less contiguous woodland may be planted: the Yoyogi woodland in Tokyo, Japan was planted in the early twentieth century, and stands at around 70 ha (Jim, 2011). The National Park of Tijuca in Rio de Janeiro (Brazil) is probably the largest planted urban woodland, at c.3300 ha (Matos et al., 2003).

All forms of urban woodland provide a range of ecosystem services, with well-documented examples including carbon storage, reduction of the heat island effect, flood attenuation, and recreational and aesthetic benefits (Turner et al., 2005; Tyrvainen et al., 2007). These services will be generally enhanced with increasing size and age of the woodland. In particular, remnant woodland may support some of the most intact soil ecosystems within the urban complex due to their lack of disturbance and management (Jim, 2010).

4.8 Rivers

As noted in Chapter 2, rivers are commonly associated with urban regions, and may be rather abundant, at least prior to their incorporation into drainage or sewerage infrastructure (often termed 'piping' or 'filling'; Wenger et al., 2009). The effects of urban hydrology on urban rivers (e.g. increased peak discharge, reduced lag times and durations of flood events, increased pollution and sediment transport, changes to water temperature and chemistry) have been noted in Chapter 3. These hydrological effects contribute to geomorphological changes such as altered sediment dynamics, changing channel morphology (Figure 4.4), reductions in geomorphic complexity, and changes to organic debris inputs; all further exacerbated by urban river management (e.g. Gurnell et al., 2007; Everard and Moggridge, 2012). Exhibition of these characteristics is generally known as the 'urban stream syndrome' (Paul and Mayer, 2001; Walsh et al., 2005). These and other effects are summarised in Table 4.1. Some of the most important ecological implications of these impacts are as follows:

1   *Reduction in diversity of most species groups.* Urban rivers are consequently ecologically poor in general, with the level of urbanisation (as measured by impervious surface cover – ISC) reducing fish and invertebrate diversity at as little as 4 per cent ISC in several studies (Paul and Meyer, 2001; Miltner et al., 2004). This trend is not universal, however. Chadwick et al. (2012), for example, found an overall positive relationship between level of urbanisation and macroinvertebrate richness in urban rivers in Florida, USA, which they attributed to increased base flows and perennial flow regimes in urban catchments which thereby allow a wider range of invertebrate species to coexist. Although unexpected, such studies help to remind us that species may respond in many varying ways to urban environments, and that we still know relatively little about urban rivers in particular

Figure 4.4 A small urban river, with the channel modified and confined by hard engineering but still showing some spontaneous vegetation growth. (Photograph by Mike Chadwick.)

Table 4.1 Overview of how urbanisation can affect river/stream ecosystems

Pathway	Effect on the system
Hydrology	Increased total runoff Decreased rise to maximum storm flows Shorter duration of peak storm flows Increased frequency of scouring flows Altered base flows
Channel geomorphology	Widened/incised channels Embedded substrates Increased scour Altered pool depths Decreased channel complexity
Water chemistry	Increased nutrients and contaminants from point and non-point sources Increased suspended sediments
Temperature	Increased stream temperatures
Trophic resources	Decreased retention of organic matter Altered organic matter budgets Altered primary production for both algae and macrophytes

Based on Paul and Meyer (2001) and Walsh *et al.* (2005).

(Wenger *et al.*, 2009; Francis, 2012). Colonisation by more urban-tolerant species may also work to obscure changes in species richness, further making generalisations difficult.

2   *Change in the composition of species assemblages.* Although diversity may not always decrease across different species groups, change in which particular species are present is common. Often species that are sensitive to urbanisation may be lost at the early stages of increase in ISC, to be replaced by less sensitive species (Walsh *et al.*, 2005). Some macroinvertebrates seem to be among the most sensitive and the earliest organisms to disappear from urban rivers (Ourso and Frenzel, 2003). In some cases the loss of diversity following urbanisation may not seem extreme, as the more sensitive species have already been lost from pre-urban rivers due to other impacts, such as those associated with agricultural land use. It is common for non-native species to occupy available niches within urban rivers (e.g. Riley *et al.*, 2005), which can lead to further impacts such as reduction in habitat quality, changes to hydrogeomorphological or biogeochemical processes, disruption of species interactions and ultimately the further decline of native species (e.g. Francis and Chadwick, 2011). Urban rivers are particularly associated with non-native and invasive species, and may represent a vector for spread throughout a region (Riley *et al.*, 2005; Säumel and Kowarik, 2010). Säumel and Kowarik (2010) demonstrated that even the seeds of tree species that are not adapted to water-based dispersal (hydrochory) can be carried at least 1200 m along an urban river within a matter of hours and potentially remain buoyant for 10 days. This illustrates the capacity for plant species to spread from urban 'source' areas via river networks, increasing rates of spread achieved by wind alone. Many aquatic non-native species, particularly invertebrates such as crustaceans, are spread around the world by hull fouling (attaching to ship hulls) or in ballast water taken on in one port and released in another (Mack, 2003). As most ports are in urban regions, they can act as points of introduction for a large number of aquatic invasives that subsequently find their way into urban river systems.

3   *Changes in system functioning, e.g. biogeochemical cycles.* Sometimes, urban rivers experience an increase in fine and coarse particulate organic matter inputs, mainly from sediment and leaf detritus arriving via runoff (Carroll and Jackson, 2009). When riparian trees and shade are removed by urban development, decreases in coarse organic matter occur while macrophytes and algae increase (Chadwick *et al.*, 2010). Organic materials may be broken down at faster rates in urban rivers due to greater flow velocities and shear stresses that physically fragment the material, increased microbial activity, or changes in the nature of the organic material itself, e.g. as non-native species may contribute leaves that decompose at a faster rate (e.g. Imberger *et al.*, 2008). However, Gessner and Chauvet (2002) have shown that changes in urban streams can result in no changes in breakdown rates because of the compensatory effect of increased nutrients and altered flow regimes. Urban rivers can also display a reduction in large wood entrainment due to a lack of riparian trees or active tree removal, which may reduce important habitat for macroinvertebrates and fish (Larson *et al.*, 2001). Recently, attempts to reinstall large wood pieces or accumulations in urban rivers have been initiated, though with little documented success (Larson *et al.*, 2001; Booth, 2005). Changes in nutrient cycling are also well documented, with

increased loadings of nitrogen and phosphorus being common. This may lead to eutrophication and the dominance of algae that reduce available oxygen and restrict photosynthesis, further impacting biodiversity and ecological quality (e.g. Wenger *et al.*, 2009). It is not just the change in volume of these inputs that is important, but also timing and duration. A river receiving elevated organic matter inputs throughout the year from increased runoff will present a modified environment to biota attuned to seasonal fluctuations in delivery, for example. All of these changes are spatially and temporally variable, which may further complicate impact and response (Wenger *et al.*, 2009), and of course functional changes are reciprocal with change in diversity and community composition, so that each influences the other.

4   *Changes in ecosystem service provision.* Rivers provide key services in the form of water and habitat provision, climate regulation and nutrient cycling, but urban rivers are particularly associated with cultural services and values. These include sport and recreation, heritage, aesthetics and spiritual/religious values. This is often reflected in the increasing recognition of the societal value of urban rivers and the explicit inclusion of social use in many urban river rehabilitation efforts (Everard and Moggridge, 2012). Urban rivers also present a particular range of disservices, including flood risk, potential health impacts, and unpleasant aesthetics or odour, all of which require management.

There is increasing interest in urban rivers, with the majority of studies focusing on the impacts of water quality and methodologies for river improvement (Francis, 2012), though many important and fundamental questions for urban river ecology remain unanswered (Wenger *et al.*, 2009). In many cases, the complexity and variability of river systems makes them difficult to generalise, and to rehabilitate. A range of urban river typologies exist across the spectrum of urbanisation, for example; Gurnell *et al.* (2007) classified 143 urban river reaches within Europe, highlighting that examples can be found of heavily engineered reinforced channels with adjacent planted trees through to well-connected and complex reaches with abundant aquatic vegetation cover. These reaches consequently present different levels of ecological quality and opportunities for ecological improvement. This is particularly important when urban rivers are considered at the landscape or catchment scale, as urban river reaches can represent gaps or filters within river networks, limiting connectivity and preventing the movement of materials and organisms, thereby impacting on species populations and increasing the risk of species loss. Most studies of urban rivers have focused on small rivers (streams) within urbanised catchments rather than large reaches within urban complexes, with a few exceptions such as the Thames in London and lower Passaic River in New Jersey (see Francis, 2012). Large urban rivers are perhaps the most deserving of future attention.

4.9 Lakes

Lakes are common features in urban areas and can provide a range of beneficial services, for example water storage. In most cities this can provide storm water protection, but in many developing areas this water can be an important source for domestic or even agricultural activities. However, the largest benefits usually come from the aesthetic value and recreation potential (e.g., swimming, boating, fishing) these systems can provide for citizens. Unfortunately, lakes suffer from altered hydrology, increased pollution and reduced ecological quality in much the same way as rivers.

Though some urban lakes are large, small and shallow lakes are more common, particularly those that are artificial and which have been created for recreation. Lakes in or near urban areas can have high levels of industrial contaminants, such as heavy metals (Baek and An, 2010). These are often concentrated in lake sediments, particularly the upper levels (Friese *et al.*, 2010). These impacts may be exacerbated by lake drainage or water abstraction, which reduces the volume of water and therefore increases the concentration of pollutants. High loads of nitrogen and phosphorus from runoff and in many cases leaky septic systems, is a major issue for urban lakes. The associated cultural eutrophication that results alters urban lake ecology dramatically. Increased nutrients promote phytoplankton, particularly cyanobacteria, and cyanobacteria blooms can be very common in urban lakes. During these blooms, toxic metabolites can be produced which are harmful to people, fish and other wildlife. Intense blooms also decrease the aesthetic value of these urban water features, both due to the thick green ooze that forms and the resulting odours associated with decomposition of organic matter. Intense blooms block sunlight and restrict photosynthesis of macrophytes. This loss of plants that stabilise banks (which can also be due to poor lake shore management) can result in erosion of lake margins and further increase turbidity. Therefore many urban lakes often display a turbid, low-oxygen state dominated by phytoplankton. The most ecologically devastating event in urban lakes is associated with high biochemical oxygen demand (BOD) of algal blooms. High BOD results in hypoxic or anoxic conditions causing fish kill and reducing food and habitat availability within the system (Poor, 2010).

Management of urban lakes requires addressing several key problems: eutrophication, toxic pollution and loss of natural shorelines. Finding solutions to restore urban lakes can be ecologically and logistically challenging. For example, algal blooms can be reduced via application herbicides, light-reducing cloths or dyes, or biological manipulations. However, these in-lake approaches only address symptoms and not causes. To truly remedy algal bloom issues, reducing nutrient load to lakes is required, but would require a whole-catchment management approach. Many toxic pollutants are found in lake sediments which can be either capped, treated or removed. These remediation techniques are again costly, but where pollutant loading has ceased, a viable solution. Shoreline degradation can be more easily approached by tree planting and landscaping; however, these activities are likely only successful when partnered with holistic and well-conserved management plans (Schueler and Simpson 2001, Carpenter and Cottingham, 1997).

4.10 Chapter summary

- Urban green space may be considered open, vegetated areas within the urban environment, and/or areas which retain features characteristic of pre-urban ecosystems. They may be abundant in urban regions, though cover can vary substantially.
- Green space consists of parks and recreational spaces, gardens, lawns, brownfield and wasteland areas, and woodland. Rivers and lakes may also be considered

forms of green space, as they are remnant ecosystems (though modified) from the pre-urban environment.

- All of these ecosystems may contain habitat for a range of species, though they vary in which species and assemblages they support, and consequently their biodiversity. Though landscape factors (e.g. position within the urban complex) may be important, the biodiversity and ecological quality of green space is usually determined by site-specific factors. Each type of green space can be quite variable in its environmental characteristics.
- Most work has been performed on parks, woodlands, gardens and rivers, with lawns, brownfield sites, lakes and allotments less well understood. Most studies are based on individual case studies rather than systematic comparisons between types of green space.

4.11 Discussion questions

1   What do you consider to be the most important urban green spaces and why? What characteristics should be used to determine their ecological value?
2   Which types of green space do you use, and why? How often do you visit urban green spaces?
3   Which types of green space are most abundant in your urban region?
4   Choose a type of green space and discuss the ecosystem services and disservices it might provide to urban inhabitants.

Chapter 5

Ecosystems within urban regions
The built environment

5.1 Introduction

The majority of the urban complex is formed from the built environment or other artificial structures, such as the varying forms of infrastructure we use for transportation or the removal of waste. At first glance, many of these structures may not appear to be particularly biodiverse or ecologically valuable, but surprising levels of diversity and some rare or unusual species and assemblages can be associated with such ecosystems. Certainly our understanding of them should be improved to allow better management of the ecological resources they represent. This chapter considers the ecology of the built environment, including installations that maintain ecological value, such as living roofs and walls, and planted trees. Terrestrial and aquatic infrastructure such as roads, pavements, railways, underground rapid transit systems, sewerage systems and canals are also considered.

5.2 Walls

Buildings and their component parts and associated constructed features, mainly consisting of walls and roofs, are integral to urban environments and comprise much of the land surface. The total 3D surface area of buildings within the urban landscape can be substantial, and building and wall structures represent extensive ecosystems that can provide habitat to a wide range of species from different taxonomic groups. Darlington (1981) estimated that around 1 ha of wall surface may exist for every 10 ha of urban land within the UK, though abundance will of course vary according to density of the built environment as well as architectural styles within a given urban region. Specific wall-associated taxa are discussed in more detail in Chapter 6 in the context of the Urban Cliff Hypothesis, but some celebrated examples of species that utilise wall habitats in the depths of urban complexes include peregrine falcons (*Falco peregrines*), buddleja (*Buddleja davidii*), bats (Chiroptera), and house sparrow (*Passer domesticus*).

Walls as ecosystems, and as a form of habitat utilised by different species, have been studied by scholars for many decades, usually due to their botanical value. The large majority of studies on walls have focused on plants. Observations have usually been made from old walls in rural or lesser urbanised areas, for example castles or fortifications (Segal, 1969; Nedelcheva and Vasileva, 2009), monuments (Zomlefer and Giannasi, 2005), walls delineating agricultural fields (e.g. Holland, 1972), and so on.

There are relatively few studies that have examined the ecology of walls in urban areas (see Francis, 2011), though significant bodies of work do exist in particular for Hong Kong (Jim, 2008; Jim and Chen, 2010) and the flood defence walls of the River Thames, London (Attrill *et al.*, 1999; Francis and Hoggart, 2012).

Although intuitively, walls are a hostile habitat and unlikely to support many species, they can be surprisingly diverse, even in the middle of heavily developed urban areas. For example, Shimwell (2009) documented 226 plant species growing on the walls of Durham in the UK, Jim and Chen (2010) found 134 plant species growing on masonry walls in Hong Kong, while Francis and Hoggart (2012) and Hoggart *et al.* (2012) recorded 90 plant species and 37 invertebrate species living on flood defence walls of the River Thames through central London (see Figure 5.1). Sustek (1999) recorded 40 species of carabid beetle around a single shop light in the centre of Bratislava over two months, many of which were utilising walls (and pavements) as habitat. In some cases, rare or endangered plant species can be found growing on walls, as observed by Guggenheim (1992) for walls in Zürich, Switzerland. Lichens have also been the focus of much attention, having distinct and diverse assemblages on wall surfaces, particularly old walls (Wheater, 2010).

Walls have all the physical requirements to support a range of organisms (and in some ways simulate cliff environments; see Section 6.2.2), and individual wall

Figure 5.1 Vegetation growing spontaneously on a flood defence wall along the River Thames through central London. (Photograph by Rob Francis.)

characteristics make a difference to the quality of the physical habitat they represent. Segal (1969) documented much of this in his book *Ecological Notes on Wall Vegetation,* and the topic was revisited by Darlington (1981) and Gilbert (1992), though there has been little further quantification since that time, with a few exceptions (e.g. Francis and Hoggart, 2012; Jim, 2008; Jim and Chen, 2010). Recently, there has been a growth of literature looking at wall ecosystems, as interest in urban ecology has developed (Lundholm, 2011; Francis, 2011). Key characteristics that influence the capacity of walls to provide habitat include wall dimensions, construction materials, inclination, microclimate, exposure, landscape position, age, moisture, sediment accumulation, pH, colour and pollution (see Wheater, 2010; Francis, 2011). It is important to recognise that walls are highly variable in their construction materials and hence their physical characteristics, so that although generalisations can be made, for example that brick and stone walls are most likely to develop fractures, interstices and rough, complex surfaces that trap plant seeds and harbour insects, much work still needs to be done in determining the importance of different wall characteristics in influencing the organisms that can live there. For example, Francis and Hoggart (2009) found that wall material type was an important driver of plant diversity along flood defence walls of the River Thames (London), but this was secondary to the disturbance caused by tidal wash within this urban estuarine system.

Often, plant species do not establish solely on the vertical wall surface, but rather on the wall top or at the base. There are consequently a series of 'microhabitats' that are found on wall ecosystems, and which relate to very fine-scale changes in water availability and sediment accumulation (Figure 5.2). These essentially consist of:

1   Wall bases, where sediment and moisture tends to accumulate, and where it is common to see plants growing.
2   Vertical wall surfaces, which display a moisture gradient as, for example, precipitation is absorbed as it runs down the surface. This often results in mosses and plants being found towards the wetter upper sections (top 30 cm) and base (bottom 10 cm) and often limited vegetation cover in between (mainly lichens), unless moisture is abundant such as along river walls or where guttering is leaking (Francis and Hoggart, 2009).
3   Horizontal surfaces such as wall tops, which also retain sediments and moisture and therefore support moss and plant species.
4   Inclined wall surfaces, which are somewhere between horizontal and vertical in their level of sediment accumulation and moisture retention, dependent on angle.
5   Vegetation growing on the wall, which may represent a form of microhabitat for other species such as invertebrates, birds and other plant species (see Wheater, 2010; Francis, 2011 for further discussion of microhabitats). Although such microhabitats are important, they may be fewer than those found on more natural rock habitats (Lundholm and Richardson, 2010).

Some distinction should also be drawn between free-standing walls, building walls and retaining walls (Lundholm, 2011). Free-standing walls are common in urban areas and may for example mark property boundaries. They are usually of limited height, relatively thin and usually constructed from wood, stone or brick. Levels of maintenance will differ depending on wall type and ownership, for example garden

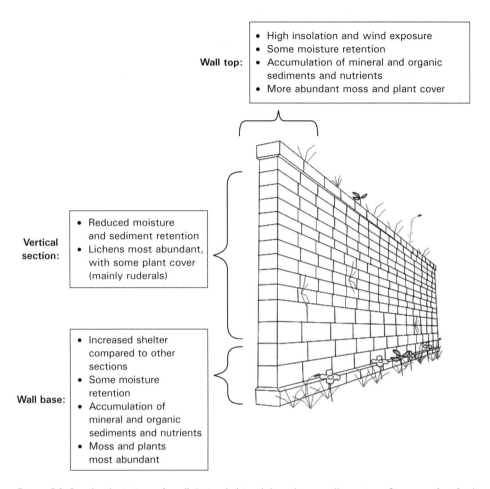

Wall top:
- High insolation and wind exposure
- Some moisture retention
- Accumulation of mineral and organic sediments and nutrients
- More abundant moss and plant cover

Vertical section:
- Reduced moisture and sediment retention
- Lichens most abundant, with some plant cover (mainly ruderals)

Wall base:
- Increased shelter compared to other sections
- Some moisture retention
- Accumulation of mineral and organic sediments and nutrients
- Moss and plants most abundant

Figure 5.2 Simple depiction of wall 'microhabitats' based on wall section. See text for further explanation.

walls may be more regularly maintained and therefore plant-free than a boundary wall along a brownfield site. Building walls will generally not support many spontaneous species due to careful design and high maintenance, unless they are abandoned, such as derelict buildings and ruins. They are also likely to have limited surface moisture due the presence of guttering to channel water away from the wall surface. Retaining walls are those constructed against a substrate, for example along a hillside to prevent slippage or along a river for flood defence. In some cases wall integrity is not essential and looser construction materials are used, such as stones without mortar. This, combined with the presence of adjacent soil that may act as a growth substrate and help regulate the wall microclimate, may support plant growth along these walls in particular, even of large trees (Jim and Chen, 2010; Francis, 2011).

Most urban wall species are generalists that survive despite the harsh conditions, rather than due to adaptations to survive in harsh conditions, though wall specialists are found in urban areas (see Chapter 6; Lundholm and Richardson, 2010; Francis and Hoggart, 2012). Of the plant species found on walls, most are small herbaceous species that complete their life cycle in one year (annuals), though longer-lived plants (perennials, including shrubs and trees) are also found. The relatively disturbed wall environment also favours disturbance-tolerating species (ruderals) rather than stress-tolerators (Francis, 2011). This may in part be a function of the way in which seeds arrive on walls; ruderal species, especially annuals, tend to produce many seeds that are wind-dispersed (anemochorous) and are therefore more likely to end up in a wall crevice. Likewise, those species eaten or otherwise associated with animals that tend to use walls (e.g. ants) are more likely to be found using wall habitat. The general pattern of colonisation for a wall is initial establishment of fungi, algae and bacteria, followed by lichens, mosses and then vascular plants of all kinds. Gilbert (1992) suggests that this process may usually take 40–80 years, assuming limited wall maintenance. Non-native species may also make up a reasonable proportion of wall assemblages, with studies demonstrating ranges between 3 and 25 per cent (Francis, 2011), though up to 91 per cent have been recorded in some cases (de Neef, 2008). Lundholm and Richardson (2010) consider that walls also act as a form of 'analogue habitat' for cliff specialists, and as such may have particular value for such species.

Although most studies have focused on non-urban walls, it is important to recognise that wall ecosystems can be important in urban regions. They are extensive in area, variable in type, and some can support a range of taxa, though they are generally species poor because they are well-maintained and colonising species are often removed to preserve the integrity of the wall structure. The growth in popularity of living walls and green facades (Figure 5.3) indicates that people can be desirous of seeing greened walls with abundant vegetation growth (as long as these are maintained; Francis and Lorimer, 2011), and so there is potential to engineer wall surfaces to support biodiversity in cities. The varying definitions of living walls are summarised in Table 5.1. Green facades and living walls are often installed for aesthetic or horticultural purposes, and have mainly been considered with respect to their ability to favourably alter urban microclimates in the same way as urban vegetation in general (e.g. Köhler, 2008). Although there are observations that green facades and living walls support fauna such as invertebrates, there has been little quantification of this (Köhler, 2008).

5.3 Roofs

Roofs also cover large areas of the urban environment; for example, Ferguson (2005) records that roof surfaces represented an average of 29–38 per cent of the 2D horizontal surface of different urban areas based on available literature. In some respects, roofs can be similar to walls in the way they function as habitat, as they are resource-poor skeletal environments that may select for or against certain species or groups. Their general horizontal arrangement can encourage species use, though there is limited investigation of roofs as habitat, with the exception of living roofs (Table 5.1). Although roofs may maintain spontaneous plant and invertebrate assemblages, most consideration of roofs has focused on bird species. Roofs are common foraging and roosting sites for urban birds, such as herring gulls (*Larus argentatus*), lesser black-backed gulls (*L. fuscus*) and rock doves (*Columba livia*), among others (Rock, 2005),

Figure 5.3 A green façade on an urban building in London. Note the bird nest boxes installed to maximise habitat use. (Photograph by Mike Chadwick.)

though these birds may also act as pests in cities if they maintain large populations (see Chapter 6). In these cases, roofs with abundant roof cavities are more supportive of bird presence and roosting. Since the mid 1990s, roofs have been extensively

Table 5.1 Definitions of terminology relating to living roofs and walls

Terminology	Definition
Green roof	Usually refers to a planted living roof and is the most common term applied, though sometimes confusion has arisen when 'green' is used in the environmental sense, such as a roof with solar panels may be considered 'green' due to its energy saving capability
Brown roof	Extensive green roofs that attempt to simulate brownfield conditions and use often use resource-poor substrates such as gravel or rubble to encourage the establishment of species that are often displaced by more competitive species
Ecoroof	An alternative for both 'green roof' and 'brown roof' generally used to refer to roofs that have been planted as extensive roofs for ecological (rather than aesthetic or recreational) purposes. Often used to avoid the idea that green roofs are in fact green and covered with lush vegetation (which many are not, but still provide habitat)
Living roof	Any vegetated roof system; a generic term for roofs designed to promote natural or planted vegetation, and used to avoid the use of the term 'green' or 'brown'. Sometimes simply referred to as a 'vegetated roof'
Intensive living/green roof	A 'roof garden' where the purpose is mainly recreation or aesthetics in the same way as an ordinary garden. Such roofs will have deeper soils, be frequently used and require regular maintenance, and may support a wide range of horticultural plants
Extensive living/green roof	A roof mainly created for supporting biodiversity or providing other environmental benefits, and which is not intended to be used frequently by humans. Usually contains a thinner layer of soil or substrate and after initial construction requires minimal maintenance and is essentially left to its own devices (including natural plant colonisation)
Green façade	Mainly refers to climbing plants such as ivy that are encouraged to grow up and along the walls of buildings (mainly on a wire or trellis framework) to form a green covering, though the roots of the plants are contained in substrate at the base of the wall
Living wall	A wall that incorporates vegetation in its structure or on its surface, and which does not require the plants to be rooted in substrate at the base of the wall as in a green façade. Most living wall systems are modular and consist of an encased growing medium placed onto the wall surface but kept separate from the wall material via a waterproof membrane, and which is watered using a drip-feed system. Natural living walls may be seen where vegetation has become integral to the wall and may help in supporting its structure, such as on some retaining walls. Sometimes living walls are bioengineered so that plant roots are purposefully used as a reinforcing mechanism within the wall structure
Biowall	A living wall or green facade that is situated indoors, primarily to enhance the atmosphere and indoor environment of, for example, shopping centres

Compiled from information found in Oberndorfer *et al.* (2007); Dunnett and Kingsbury (2008); Emilsson (2008); and Köhler (2008).

Reproduced from Francis and Lorimer (2011), with permission.

enhanced in some areas via the installation of living roof structures (e.g. Oberndorfer *et al.*, 2007; Francis and Lorimer, 2011).

Living roofs (Figure 5.4; see Table 5.1) are frequently retrofitted to existing buildings, or increasingly incorporated into new builds. They usually consist of a root-resistant membrane surmounted by drainage layers, a filter membrane and a mineral substrate that may vary in depth, composition and organic content. They are typically planted with species that can tolerate relatively harsh, xeric conditions such as *Sedum* spp. (Emilsson, 2008), or sometimes seeded with a mix of herbaceous species, but may sometimes be left to colonise spontaneously. As with living walls, most studies of living roofs have concentrated on their benefits to urban microclimate and water management at varying spatial scales, though there has recently been an increasing focus on the ways in which they can support different species (Oberndorfer *et al.*, 2007; Dunnett *et al.*, 2008; Francis and Lorimer, 2011). Usually roofs installed to support urban biodiversity have intentionally simulated brownfield sites, which have been demonstrated to have relatively high ecological quality within an urban context, at least for some taxa (Schadek *et al.*, 2009; see Section 4.6). The potential may exist to simulate other forms of habitat by modifying substrate characteristics such as construction material, organic content and pH, and thereby support different species and assemblages, but this form of ecological engineering remains in its infancy.

Figure 5.4 A living roof installed on a building in Birmingham, UK. This is an 'extensive' roof, created to simulate brownfield habitat. (Photograph by Adam Bates, used with permission.)

Many urban regions have incorporated living roofs and walls and there has been notable expansion in recent years. Some key examples include London (Grant, 2006), Chicago (Yang *et al.*, 2008), Washington, DC (Niu *et al.*, 2010), Shanghai (Roehr and Kong, 2010), New York City (Gaffin *et al.*, 2009), Singapore (Yuen and Hien, 2005) and Toronto (Roehr and Kong, 2010), among others.

Substrate conditions and initial planting have an important influence on the diversity and types of species found on living roofs. Most roofs are designed based on principles laid out in the Forschungsgesellschaft Landschaftsentwicklung Landschaftsbau Guidelines (FLL, 2002) and its various interpretations. Various combinations of substrate depth, substrate materials, organic content, moisture, pH, nutrient levels and initial plantings are achievable, which will influence the types of species and assemblages that will emerge on living roofs (e.g. Emilsson, 2008; Olly *et al.*, 2011). Other important factors determining both diversity and assemblage composition include: (1) roof area, with larger roofs supporting more species (Oberndorfer *et al.*, 2007); (2) roof age, with older roofs having different more developed soils, for example with lower pH and higher nutrient levels, alongside greater time for species colonisation and biotic interactions (e.g. Schrader and Böning, 2006; Rowe *et al.*, 2012); and (3) habitat heterogeneity or provision of microsites such as patches of different substrate type, wood or piles of stones on the roofs that may support a range of different taxa (e.g. Bates *et al.*, 2009). Roofs are most likely to benefit organisms of relatively small body size and with high capacity to move and disperse through the urban environment, such as flying insects and birds. Diversities of such species can be high, with hundreds of species being found on individual roofs (e.g. Kadas, 2006). Species that are not mobile or do not disperse well are often rare or scarce on roofs (Francis and Lorimer, 2011). Those plant species that are dispersed via anemochory or zoochory are more likely to be deposited on roofs than species dispersed via other means, such as hydrochory, or requiring specific treatments (e.g. burial) before germinating. Living roofs nevertheless present opportunities for urban ecological improvement following the philosophy of reconciliation ecology (see Chapter 7).

5.4 Transport infrastructure

5.4.1 Roads

Although transport infrastructure such as roads and railways is put in place specifically to facilitate the movement of humans around urban regions, such transport corridors can act as habitat as well as facilitate species movement (Forman *et al.*, 2003; Hayasaka *et al.*, 2012; cf. Kalwij *et al.*, 2008). Most studies into the ecology of roads have understandably focused on exurban and non-urban roads, which cut across a range of natural and semi-natural ecosystems and can cover vast areas, with some large countries (such as the USA) having millions of kilometres of road, of which around 80 per cent are in non-urban areas (Forman *et al.*, 2003). Because urban roads in particular are frequently disturbed and heavily utilised, they may not be as diverse as in non-urban areas, but nevertheless can support a selection of species and have their own ecology.

Roads are ubiquitous within urban areas (Forman *et al.*, 2003), and although their spatial density and total surface area vary depending upon factors such as housing density and urban land use (Arnold and Gibbons, 1996; Hawbaker *et al.*, 2005), they are dominant features of the urban complex. For example, Arnold and Gibbons (1996) note that roads covered between 25 and 60 per cent of the urban land surface in Olympia, Washington (USA), depending on the density of buildings in various urban sectors. Roads are landscape features that many people interact with on a daily basis, and can act as conduits, barriers and filters to species, depending upon road width, density and construction materials, and species characteristics such as size, vagility and behaviour (Forman, 1996; Forman *et al.*, 2003; Parris, 2006). Urban roads may: (1) provide a limited form of habitat for species, particularly plants and invertebrates; (2) increase the relative isolation of other urban ecosystems and species populations, dependent on the capability of the species to disperse across road barriers (Parris, 2006); (3) increase mortality of fauna (Forman, 1996; Forman *et al.*, 2003; Orłowski, 2008); and (4) facilitate the spread of both native and non-native species (Hansen and Clevenger, 2005). As Forman *et al.* (2003, p. 90) notes, 'there is nothing in nature that mimics a road', and they are therefore good candidates for 'novel ecosystems' (see Hobbs *et al.*, 2006; Section 6.4).

It is primarily roadside verges (Figure 5.4) that offer the greatest opportunity for species to utilise these corridors as habitat, though such verges may be lacking in some urban areas. Verges in less dense urban areas may consist of sown grassland or wild flower assemblages, trees or even sedum mats (Forman *et al.*, 2003; Dunnett and Kingsbury, 2008), though verges also experience colonisation from plants within the surrounding environment, sometimes transported by animals moving along verges or by vehicles. Vegetation can be diverse (Hayasaka *et al.*, 2012) and support a range of invertebrates (Saarinen *et al.*, 2005); though urban road verges (where they exist) are necessarily more limited than those in rural areas, and may be more frequently and intensively managed (Saarinen *et al.*, 2005). In some cases, managed verges may be comparable to lawns, while sometimes spontaneous colonisation may be observed, usually of generalist light-demanding species, though wetland species may also be found along roads where they maintain adjacent ditches (Forman *et al.*, 2003). Verge plant diversity is likely to increase in exurban areas compared to the urban complex. Plant diversity can be relatively high, and several studies have found that exurban roadsides may maintain a significant proportion of regional plant diversity, e.g. 44 per cent of the UK and 50 per cent of the Netherlands (Forman *et al.*, 2003). Though local diversity can be high, the high connectivity of road networks and the dominance of generalist species that are easily dispersed may lead to homogenisation of assemblages at the regional scale (Hayasaka *et al.*, 2012).

Relatively high species diversity and density may be found in part because road verges are 'ecotonal' habitat representing a transition from the disturbed resource-poor 'skeletal' habitat of the road surface to whatever the adjacent land use is – as such they demonstrate characteristics typical of 'edge' habitat, including relatively high diversity over small spatial scales, an abundance of generalist species, and an accumulation of sediments, nutrients, pollutants, insects and seeds (see Box 2.3). Roadside verges are also hotspots for the accumulation of de-icing salts in particular, and as such may support higher than predicted levels of salt-tolerant species, including non-native and invasive species (Šerá, 2008). Verge width and height, as well as land use, may influence the species found, with larger verges supporting more species and taller vegetation or nearby woodland providing useful structural diversity and

Figure 5.5 A managed urban roadside verge of grasses, herbs and trees.

source populations (e.g. Saarinen *et al.*, 2005). Patches of vegetation at road intersections have also been shown to support plant and invertebrate species associated with grasslands, though not to the same extent as some urban green spaces (Valtonen *et al.*, 2007). In urban areas, intersections and roundabouts often have a vegetated centre that may act as habitat for some species (e.g. Helden and Leather, 2004). In some cases, roads are heavily managed and apart from occasional planted trees, do not support many species.

Infrastructural surfaces (tarmac/asphalt, concrete, paving slabs) can also provide habitat to species in a similar way to walls. They offer harsh conditions, but are perhaps in some ways less harsh than wall surfaces as they are horizontal and are more likely to retain sediment and moisture, at least in the case of paving tiles or where surface cracks are present. Concrete and tarmac/asphalt offer fewer opportunities for species to establish than paving due to more resistant surfaces and generally higher frequency of disturbance. With the exception of trees, which are planted in specially constructed pits along infrastructure (and are discussed in Section 5.4.3), most plant species found on paved surfaces are common generalists with some rock specialists, as for walls (Wheater, 2010). Pavement often does not contain mortar, and so it may be easier for plant seeds to establish in interstices, though plants still have to overcome the obstacles of poor, compacted soils and limited rooting space. They may also experience the extra disturbance of trampling and regular management such as weeding or application of herbicides, further restricting the species that may be found on pavements and selecting for species tolerant to such disturbances (Wheater, 2010).

Roads, and urban roads in particular, remain an important area for further research (Forman *et al.*, 2003). The Rauischholzhausen Agenda (Roedenbeck *et al.*, 2007) identified that the impact of road networks on species populations at the landscape scale needs to be more thoroughly evaluated, and this also holds true for any benefits associated with road habitats, such as verges. It is not well established to what extent urban roadsides act as population sinks, whether species populations remain viable over time, or whether they function well as habitat or movement corridors within the urban complex. Given their abundance in the urban region, understanding their ecology is important in determining how they might best be managed for biodiversity and ecosystems services.

5.4.2 Railways

Railways are also an important form of urban infrastructure. Like roads, they are extensive networks that run through urban regions, but as they are usually composed of railings on beds of crushed rock and gravel, they offer a habitat similar to scree, and like many other skeletal urban ecosystems may support abundant generalist species as well as occasional rock specialists. Railway verges may be similar to roadside verges and have abundant herb and grass species, and relatively high diversity; for example, Brandes (2005) records over 1000 plant species associated with urban railways in Germany. Wahlbrink and Zucchi (1994) found 52 carabid beetle species along urban railway embankments of Osnabruck, Germany, with a reduction of both diversity and individual body size with proximity to the urban centre, along with changes in assemblage composition. As landscape barriers, railways are often narrower and therefore offer less resistance to species movement than other forms of urban infrastructure (Tremblay and St. Clair, 2009). Railways are commonly associated with non-native species (Brandes, 2005); of the 1000+ species recorded from railways in Germany by Brandes (2005), 309 were non-native. Some invasive species are particularly able to exploit the rock-like habitats provided by railways. For example, *Buddleja davidii* is particularly common on railway gravels and can frequently be seen growing on walls and bridges along train lines, particularly in Europe (Ebeling *et al.*, 2008).

Railways may in some cases act as conduits for species dispersal. Penone *et al.* (2012) observed that railways were landscape corridors for some grassland plant species in the Paris region. They may in particular act as vectors for the dispersal and spread of non-native and invasive species (e.g. Hansen and Clevenger, 2005), with *Buddleja davdii* again being a notorious example, though Penone *et al.* (2012) found that the presence of invasives was more associated with general urbanisation rather than specifically with railways. Nevertheless, their contiguous nature means that they may allow for the movement of organisms within and between regions.

Underground rapid transit (rail) systems are common in many large urban complexes. Such underground networks present different environmental conditions to aboveground urban areas, and consequently support distinct assemblages and sometimes harbour pest species that have significant human impacts. In some cases, underground species populations may become quite distinct. The London underground mosquito (*Culex pipiens* f. *molestus*) for example is a genetically distinct form or biotype (see Byrne and Nichols (1999) for a discussion of classification) of the common house mosquito *Culex pipiens* that seems to have colonised the underground network at some point prior to the Second World War (1939–45), and has since become a biting nuisance for humans on the underground system, as well as a potential disease vector for West Nile virus and dengue fever (Byrne and Nichols, 1999), though the risk of epidemics resulting from the species is considered low in many regions at the present time (Higgs *et al.*, 2004). This species has also been found in the New York and Amsterdam metros and subterranean Chicago, and essentially represents an indoor form of the species (Higgs *et al.*, 2004). In many cases, underground populations of the species are subject to control as a result of their pest status. Little is known of the ecology of underground transport systems in general, and many interesting questions remain unanswered.

5.4.3 Infrastructural trees

Individual trees are perhaps the most common ecosystems found within urban areas, and collectively represent a substantial resource for habitat and ecosystem services as with all urban vegetation, though trees in particular are highly visible and consequently may capture the popular imagination and garner public support more than some less obvious greenery (Adams, 2005). Trees growing spontaneously or having been planted in urban parks or gardens have been included in the broader discussions of these ecosystems in Chapter 4; here, trees growing alongside urban infrastructure are considered. Within urban infrastructure, trees are the most commonly planted type of vegetation (e.g. many thousands may be present in individual cities; Jim and Chen, 2008), and are often found along roadsides, streets, pedestrian areas, canals and riversides (Figure 5.6). In an extensive survey of street trees in various European urban regions, Pauleit *et al.* (2002) found that most regions had between 50 and 80 trees per 1000 residents, with a range of 20–140+ trees per 1000 residents. This may reflect the relative age and less-dense structure of many European cities; in the much younger and denser Hong Kong region, around 6 trees were found per 1000 residents by Jim (2001). The abundances of infrastructural trees will vary based on many factors, including the age of the urban region, the cultural history of the region (e.g. whether trees were widely planted in the past or not) and current cultural perceptions of the trees (e.g. whether they are perceived as a good thing or as a risk to people or infrastructure). Infrastructural trees may also present a range of disservices, including damage to infrastructure or buildings from root or branch growth, obstruction of views, spread of plant diseases, contribution of allergens to the atmosphere, and collapse onto people or property. Dawe (2010) argues that due to increasing perception of such threats associated with trees, some urban regions are losing many of their infrastructural trees; or, where numbers are being maintained by active planting programmes, this is at the expense of the loss of many of the more mature trees that may have been present from a less urbanised period. Such trees are ecologically important for the types of habitat they may represent, or as species that are relatively rare within either an urban or a wider context, but are also the trees that are more likely to present a potential hazard to the public or infrastructure.

The majority of trees are planted in harsh conditions, for example in poor soils (see Section 3.7) or with limited rooting space, in polluted environments, in high-temperature and low-moisture microclimates, with insufficient natural light or disruptive artificial light, or subject to water stress, which reduces the growth and lifespan of some species and increases mortality rates (e.g. Pauleit *et al.*, 2002; Sæbø *et al.*,

Figure 5.6 Roadside trees lining the Embankment next to the River Thames in London. The trees are London plane (*Platanus × acerifolia*), which are frequently planted and common throughout the London area.

2003; Bühler *et al.*, 2009). Pauleit *et al.* (2002) note recorded average life span estimates of 10–60 years for street trees, though of course much depends on planting conditions and species under consideration. In particular, the volume and depth of the initial planting pit for street and roadside trees seems to strongly affect growth and survival (Bühler *et al.*, 2009). There is also now a greater attempt to provide 'structural soils' that contain a significant proportion of coarse sediment (gravel and larger) to take some of the weight of the urban surface and prevent soil compaction, while still allowing sufficient fine sediment for root growth. Planted street trees are often frequently and regularly irrigated as part of management and maintenance – irrigation and sufficient water supply is particularly important when trees are establishing, but may artificially increase water availability beyond 'normal' levels, restricting root growth and making them less likely to perform well and survive in the long term (Bühler *et al.*, 2009). Indeed, street trees need to be managed throughout their lifetime, a commitment that may also lead to poor growth and high mortality when not followed through.

Alongside poor growth conditions, disturbances to trees are also common, through construction work, underground placement of utilities or vandalism (e.g. Alvey, 2006; Bühler *et al.*, 2009). Pauleit *et al.* (2002) found that up to 30 per cent of newly planted trees experienced vandalism in some urban areas (e.g. the UK), though levels were below 5 per cent in central Europe. Trees are also limited in the ways in which they may grow due to the requirement not to obstruct or interfere with infrastructure, whether physically or by, for example, obscuring street signs or blocking lines of sight

(Jim, 2001). Species are selected primarily to cope with difficult growing conditions alongside culturally desirable attributes such as large height and crown spread to provide shade or shelter, aesthetic appearance, and ease of maintenance (e.g. Sæbø et al., 2003). Consequently in many urban regions such trees are selected from only a few different genera or species, some of which are cultivated for the purpose (Pauleit et al., 2002; Sæbø et al., 2003; Alvey, 2006). Pauleit et al. (2002) found that 50–70 per cent of street trees planted in Europe came from only 3–5 genera, though a wider range of species was planted in general in southern Europe. Sæbø et al. (2003) recorded that between 30 and 90 per cent of trees planted in paved areas came from one or two species in urban regions of Iceland, Norway and Finland.

Despite this, the overall diversity of infrastructural trees within an urban region can be quite high. Jim (2001) found 149 species along roads in the very dense and poorly greened urban region of Hong Kong, though the majority of the trees (55.7 per cent) were composed of just 10 species. Diversity varies widely, however; Jim and Chen (2008) found only 40 species of street trees in a survey of Taipei, Taiwan, from over 8000 individuals. Subburayalu and Sydnor (2012) recorded 58, 102 and 170 species of street trees within three urban regions within Ohio, from samples of 18,662, 12,176 and 84,782 individuals, respectively. Sjöman et al. (2012) found a range of 24–113 species of street tree in 10 urban regions of Norway, though higher diversity was often associated with larger numbers of non-native planted trees. Many factors influence the diversity of trees found, though the key ones relate to planting and management strategies. Non-native species are often planted (Bühler et al., 2009), and historically this has been performed with little consideration as to the potential threat such species may pose to other ecosystems and the wider region should they become 'problem species' (i.e. invasive). Nevertheless, it is recommended that a wide diversity of street trees is planted in order to both provide a range of habitat and reduce disservices such as the spread of tree disease, which may often be species-specific (e.g. Dutch elm disease or ash dieback; see Bühler et al., 2009). The approach in many cases has been to select trees for difficult growth conditions, rather than improving, e.g. soils to allow for a wider range of species to be planted (Jim, 2001; Jim and Chen, 2008). Sæbø et al. (2003) recommend that species selection should be based on capacity to grow in a range of climatic envelopes, to resist disease, to maintain a long life span, and to grow well in a range of environmental conditions in order to maximise success. Quigley (2004) notes that trees found in the early to middle stages of succession, and which are therefore more adapted to environmental stress and disturbance, perform better as street trees.

Infrastructural trees are important not just in themselves, but also as habitat for other species. Alongside the many invertebrates that utilise urban trees as habitat, the planting of certain species may particularly increase populations or ranges of birds and mammals. Populations of the grey-headed flying-fox (Pteropus poliocephalus) have increased substantially in Melbourne, Australia, since the planting of many trees (95 species, and a conservative estimate of 315,719 individuals) that may provide food for the species. Soh et al. (2002) found that the planting of yellow flame trees (Peltophorum pterocarpum) was important in maintaining populations of the invasive house crow (Corvus splendens) in Singapore, due to their suitability as nesting sites (see Box 6.1). Whether the species they support are considered beneficial or problematic, infrastructural trees are abundant in urban regions, and represent an important component of the urban forest.

5.4.4 Canals

Despite the sometimes extensive networks of canals that run through some urban regions, there has been limited ecological investigation of these urban ecosystems. Although in some ways similar to urban rivers, canals are often artificially constructed or are very heavily modified river channels, and are purposely designed for transportation or irrigation/drinking water supply. Their hydrology is entirely regulated and as such they experience very little dynamism. The large majority are freshwater, though large coastal or estuarine ship canals leading to inland ports are common and in some regions coastal canals can be found (the most famous example being Venice, Italy, though in this case the city is constructed on coastal islands and so the canals are not comparable to other urban regions). Canal networks can be extensive (for example around 3000 km of canals exist in the UK, though only a limited proportion is in urban areas; Faiers and Bailey, 2005). Canals can support a range of aquatic taxa, including plants, diatoms, algae, macroinvertebrates, fish, mammals and birds (Dorotovicova and Ot'ahel'ova, 2008). Their towpaths are often vegetated (Figure 5.7), and may for example maintain hedgerows or scrub that may act as habitat for a variety of species (e.g. Faiers and Bailey, 2005). Canals are often heavily polluted, and this can influence the diversity and composition of canal assemblages, though impacts vary depending on level of pollution and the frequency, intensity and duration of urban water inputs, whether from surface runoff or sewerage system overflow (Dorotovicova and Ot'ahel'ova, 2008). Sewage contaminants in urban canals often lead to high levels of viruses and bacteria that may be pathogenic to humans (Reuben *et al.*, 2012).

Figure 5.7 An urban canal. Even in the middle of an urban complex, canals can provide long strips of dense vegetation along their towpaths, as well as representing aquatic habitat. (Photograph by Rob Francis.)

5.5 Sewerage systems

Extensive networks of sewers exist under urban complexes, sometimes incorporating small urban rivers that have been entirely converted to enclosed channels. Although generally inhospitable to most species, assemblages (mainly composed of microbial and invertebrate species) nevertheless exist in these subterranean landscapes and may be entirely novel. McLellan *et al.* (2010) found a unique microbial assemblage from sewage in urban Milwaukee, composed of those associated with human faeces and the wider environment in general. These microbial communities in particular were influenced by the hydrology of the sewerage systems, with rainwater and stormwater inputs changing microbial populations and community structure, as, for example, bacteria from urban surfaces and soil were washed into the sewerage systems. Certain bacteria are particularly abundant in sewerage systems – examples include sulphur-oxidising bacteria such as *Thiobacillus* spp., which produce sulphuric acid and deposit it on pipe surfaces (Okabe *et al.*, 2007). This has been particularly notorious for corroding both concrete and steel sewer pipes, to the extent that alternative materials have been developed for such infrastructure (Haile and Nakhla, 2008).

Perhaps the species most commonly associated with sewers are ubiquitous urban pests that use the sewers as relatively safe, sheltered environments that may have abundant food resources from human domestic waste, or offer direct access to human homes. Key examples include cockroaches, brown rats (*Rattus norvegicus*), and mosquitoes, and sewerage systems may represent 'source' ecosystems with a population surfeit, supplying individuals to aboveground systems where they have their greatest impact (e.g. Gras *et al.*, 2012). However, more unusual species can be found in urban sewers, and the number of species utilising these habitats is likely to increase. Csuzdi *et al.* (2008) describe how an opportunistic earthworm native to eastern Africa, *Dichogster bolaui*, escaped from greenhouses to colonise sewerage systems and establish breeding populations in both Hungary and Israel. This is the first earthworm to be known to have 'naturalised' within sewerage systems (and from there to residences), and suggests that sewerage systems may represent suitable habitat for some tropical species within temperate urban environments due to warmer temperatures (from heating and warm water) and abundant moisture (Csuzdi *et al.*, 2008). Despite not being pleasant ecosystems to work in, sewerage systems can be very interesting ecologically.

5.6 Chapter summary

- The built environment is composed of varying types of urban ecosystem, including wall and roof components of buildings, and infrastructure such as roads, railways, canals and sewerage systems.
- Walls are abundant in the urban environment but have mainly been considered in non-urban contexts. They have potential for ecological engineering, but this remains at the early stages of investigation.
- Roofs are also abundant and under-researched, though the installation of living roofs is a growing phenomenon. The ecological value of living roof installations depends largely on their design, and again this is only partly understood.

- Roads and railway verges may provide habitat for species, and may also be routes of spread for species through the urban landscape.
- Infrastructural trees are widely planted and can be highly valued, though tend to be drawn from a limited number of species and are often planted in conditions that are not ideal for growth or longevity. There is now greater consideration of species selection and planting techniques in urban areas.
- Canals and sewerage systems represent under-explored but novel ecosystems. Though usually associated with pollution and pests, they can maintain interesting environmental conditions and species assemblages.

5.7 Discussion questions

1   What are the main similarities and differences between urban green spaces and the built environment?
2   Have you seen a living roof or wall? Why is their popularity growing in urban areas?
3   What factors would you need to consider when planning a planting campaign for infrastructural trees?
4   Why are sewerage and canal systems relatively under-explored as ecosystems?

Chapter 6

Urban species

6.1 Introduction

The previous chapters have established broad trends in biodiversity within urban regions, and have described the key ecosystems nested within urban regions. This chapter now considers in more detail the types of species found in urban regions. It does not consider individual species, except as case studies or examples, but rather key characteristics associated with urban plants and animals. The chapter also considers the many ways species may respond to urban environments. There has been limited investigation of patterns and dynamics of different species categories in urban environments relative to non-urbanised areas, though some general trends have been observed and are well accepted. For example, the dominance of generalist species and an elevated proportion of non-native species are consistently found in many urban ecosystems. These and other aspects of urban species are considered below.

6.2 Generalists vs. specialists

One relatively consistent trend is the prevalence of generalist species in urban ecosystems. This pattern is also associated with both declines and/or increases in the relative abundance of specialist species (Jokimaki *et al.*, 2011). A generalist is usually considered to be a species that demonstrates flexibility of behaviour, morphology, growth form, physiology, etc., or has a wide tolerance for different environmental conditions, that allows it to utilise a range of environmental resources and therefore persist in varying environments. These characteristics are often combined with high rates of reproduction and short life spans leading to high population growth rates and the capacity to cope with disturbance and changing environments in many cases. Note, however, that this is only a broad interpretation: generalists may be considered in terms of niche breadth (capacity to use a range of resources as described above) and niche positioning (the role or 'position' of a species within a community, and how 'typical' the resources it uses are for those available within that ecosystem (see Evans *et al.*, 2011). Most urban studies either do not consider any distinction between the two, or focus on the former, though notable differences can be found depending on how 'generalists' are defined (Evans *et al.*, 2011). Here, the looser 'niche breadth' interpretation is generally followed, in line with the majority of studies. A specialist species has more specific and therefore constrained (in space and/or time) environmental requirements and limited niche availability, and therefore they generally have more limited distributions and lower populations, apart from where conditions are optimal.

Many urban species are generalists that can utilise urban resources as well as those in other ecosystems. These species generally maintain both relatively wide distributions and high populations (e.g. McIntyre, 2000). The trend for increasing distributions, abundance and diversity of generalist species has been demonstrated for many species groups, including plants (Van der Veken *et al.*, 2004), birds (e.g. Møller, 2009; Carbó-Ramírez and Zuria, 2011; see also Evans *et al.*, 2011), butterflies (Lizée *et al.*, 2011) and beetles (Tóthmérész *et al.*, 2011). For example, ruderal plant species, which are those that have short life spans, prolific seed production and which have evolved in disturbed environments (Grime, 1977) are particularly diverse and abundant in urban ecosystems (e.g. Albrecht *et al.*, 2011; Bigirimana *et al.*, 2011). Animals that exhibit flexibility of resource use, e.g. nest or den sites, foraging activities, food sources and so on, are also relatively successful in urban environments, and can have high abundances (Jokimaki *et al.*, 2011; see Section 6.8). Increases in generalists are linked to the common urban characteristics of increased resource availability, high productivity and somewhat stable climatic envelopes (Shochat *et al.*, 2006), which create conditions that support species with flexibility in their resource use and/or competitive advantages based on those resources that are abundant (e.g. anthropogenic food waste).

The increase in generalists is usually associated with a decline in specialist species. Such declines are increasingly notable along gradients from green space to the built environment. This has been demonstrated in particular for woodland or forest species, and may be exhibited at a range of spatial scales. At local scales, such as in and around a patch of green space, abundance, proportion or richness of specialists is likely to be associated with the centre of a patch compared to the edges or the surrounding matrix (see Chapter 4). This is particularly true for large patches, and is likely to reflect the preservation of interior habitat that some specialists will require (Carbó-Ramírez and Zuria, 2011). Conversely, in smaller patches specialist species are likely to have low populations or be absent altogether. At regional scales, declines in specialist species have been observed with increasing proximity to urban centres, generally reflecting the decline in abundance, patch size and contiguity of green space observed as the built environment becomes older and denser with increasing centrality, compared to the exurban land (Lizée *et al.*, 2011). These two interacting gradients may create a complex mosaic of species distributions within the urban region, with general declines towards the centre of the urban complex interrupted by relative 'peaks' of specialist populations or diversity in embedded green spaces.

These trends do of course depend on the species and the type of green patch in question. Some specialists may actually increase in urban environments, such as those that are naturally found in rock/cliff ecosystems, which may be particularly supported by the built environment in the same way as generalists (see Chapter 5; Section 6.2.2; Larson *et al.*, 2004; Lundholm and Richardson, 2010).

The ways in which the spatial configuration of urban landscapes may influence patterns of diversity have been discussed in Chapter 2. In terms of the persistence of generalists and specialists within the urban complex, two further concepts are also important to consider: the mass effect and the urban cliff effect.

6.2.1 Mass effect

One explanation for the abundance and high diversity of generalist species (in particular ruderal plant species) within the urban environment is the mass effect (sensu Shmida and Ellner, 1984), or 'source–sink' dynamics (see Section 2.5.1). This is essentially where some ecosystems support populations that are very successful ('source' areas) and consequently 'export' individuals to areas that are less favourable, and where mortality may exceed reproduction (termed 'sink' areas). In this sense, the poorer 'sink' areas may display a high level of diversity, but this is essentially artificial, maintained by the success of the 'source' areas via a 'rescue effect' of repeated colonisation. Much landscape ecology theory has evaluated this, and the trend is common in fragmented landscapes (e.g. Lobel *et al.*, 2006). Within this context, the loss of a source area has a significant impact on species with spatially segregated distributions within a region.

   Urban ecosystems may be a good example of this at both regional and local scales; organisms from natural or semi-natural 'source' ecosystems outside the urban complex may support urban populations via such a rescue effect. Likewise, the diversity of harsher artificial 'sink' ecosystems may be maintained by a rescue effect from more successful ecosystems such as parkland or gardens. Several studies have found that urban assemblages on such artificial substrates are composed of generalist species from the surrounding ecosystems, and are probably maintained by a spatial mass effect (e.g. Francis and Hoggart, 2012), while Vierling (2000) determined that urbanised areas acted as sinks for red-winged blackbird (*Agelaius phoeniceus*) populations probably due to an abundance of synathropic predators. The potential dominance of the mass effect echoes a trend found globally for areas of high population to maintain those where populations are in decline or extirpated, and may be partly responsible for temporary maintenance of biodiversity in areas where populations are not viable in the long term (the 'extinction time lag'), which often occurs following habitat degradation and loss. Loss in urban biodiversity over time may be further offset to some extent by the arrival of non-native species (see Section 6.3.1), some of which may also be existing on the 'edge' of their niches.

6.2.2 Urban cliff effect

One group of specialist organisms that may experience an increase in range or populations due to urban ecosystems is those associated with cliff, rock or talus (scree) habitat. Larson *et al.* (2004) argued that the built environment is essentially a more sophisticated replication of cave and rock-based dwellings that humans would have used as an early form of habitation. As a result, artificial niches for such species have been created in urban environments, allowing them to colonise and persist (termed the *urban cliff effect* or *urban cliff hypothesis*). Indeed, many species that are associated with agriculture, including weeds, may have originated in cliff ecosystems (Larson *et al.*, 2004). As a result, humans have both intentionally and unintentionally spread rock-based species and even entire communities around the world. There is certainly evidence that, although the majority of species in urban environments are not rock specialists, the number of species associated with cliff or rock ecosystems is higher than would be expected if urban species were drawn from an essentially random pool of natural ecosystems. Francis and Hoggart (2012) found that 16 per cent of wall vegetation found along the tidal Thames through central London was typical of cliff/rock ecosystems, which was comparable to the number originating from grassland (17 per cent) and woodland (15 per cent) ecosystems. Albrecht *et al.* (2011) found that seed banks in urban railway sidings were mainly composed of ruderal species and those associated with rocky habitats, though they did not always manifest aboveground due to competition from woody species. Lundholm and Richardson (2010) offer several other clear examples of rock/cliff species that thrive in the built environment, from obvious examples such as the rock dove or pigeon to scorpions. Given that cliff/rock ecosystems only account for around 0.5 per cent of the global surface (Larson *et al.*, 2004), an urban cliff effect is clearly occurring. In some cases, those species that are considered 'generalist' ruderal species are naturally (or were originally) associated with disturbed rock-based habitats (Larson *et al.*, 2004), meaning that the urban cliff effect may be more extensive than initially supposed.

6.3 Pre-urban vs. colonising species

Most urban ecosystems will contain a mix of species that existed prior to urbanisation (pre-urban), along with some that have colonised since urbanisation (colonisers). Some pre-urban species are remnant, with limited distributions and populations, and may often be habitat specialists (e.g. woodland or wetland specialists) that cannot cope well in the general urban environment. Some common urban species will be those found in the pre-urban environment that have nevertheless responded well to the resources provided by urbanisation, and have subsequently expanded in range and population.

However, many species found in urban environments are new colonisers, i.e. have arrived following urbanisation. There is little information and comparison on the assemblages present prior to and following urbanisation, as little before/after research has and can be done. In many ways, as urban ecosystems have no 'typical' communities, the question has been seen to have limited relevance (e.g. Hinchliffe *et al.*, 2005). However, much interest has been generated in the abundance of non-native species that have colonised urban environments, which are not only new to the urban region, but to the wider country or biogeographic region. Non-native species, and the problematic subset often termed 'invasive' species, are closely associated with urban environments.

6.3.1 Non-native and invasive species

Non-native or 'alien' species may be defined in different ways, but are most simply considered to be those species that are moved outside their natural range ('introduced') by human activity. Non-native species that are introduced and subsequently establish, spread effectively and have a detrimental effect on their host ecosystem, are often classified as 'invasive' or 'invasive alien species'. Invasive species have been regarded as one of the greatest threats to current global biodiversity (McGeoch *et al.*, 2010). There is some disagreement on the definitions of 'invasive', including whether 'invasive' species should refer only to those that spread rapidly rather than those that have a detrimental effect, or do both, and also whether they automatically cover non-native species or not (see Ricciardi and Cohen, 2007). Some native species may display rapid invasion and displace other species causing a reduction in diversity in a localised area, for example. Usually the term 'invasive alien species' or 'IAS' is used to clarify the distinction between native and non-native, though the question of rate of spread vs. detrimental effect is more open to interpretation. Here we use 'invasive' to refer to non-native ('alien') species that cause a notable level of ecological, environmental, economic or societal harm. The IAS threat is almost entirely catalysed by humans, who have been responsible for the intentional and unintentional introduction of species to unfamiliar ecosystems for millennia (Elton, 1958; Vitousek *et al.*, 1997; Rackham, 2000). However, it is only recently that this process has been recognised as a problem, at first due to the threat they may represent to native biotic assemblages (e.g. Elton, 1958) and subsequently (particularly from the late 1980s onwards) mainly due to unintended consequences of introductions that impact upon economic activities within the host system (e.g. crop herbivory, damage to infrastructure) or quality of life (health impacts, loss of charismatic or recreationally important species, compromising of ecosystem services) (Pimentel *et al.*, 2005). Only a small proportion of non-native species become invasive, however, and the large majority do not have substantial impacts on their recipient environment or species assemblages.

Urban regions represent a major point of both intentional and accidental introduction of non-native and invasive species; for example the planting of non-natives in gardens or parks (Smith *et al.*, 2006a), and arrival on ships in ports (Clark and Johnston, 2009). Many species are imported specifically for horticulture, aquaculture and the pet trade, while others arrive as 'passengers' on other species, including plant seeds, pests and pathogens. Aquatic introductions may be particularly notable in coastal urban regions. Many contain ports and are a focus for international shipping, leading to accidental introductions of both marine and freshwater non-native species via hull fouling and ballast release (Mack, 2003). Many thousands of species (7000–10,000; Mack, 2003) are moved around the global oceans daily in commercial ships, and shipping has probably been responsible for some of the most problematic aquatic introductions such as the Chinese mitten crab (*Eriocheir sinensis*; Bentley, 2012) and Asian clam (*Potamocorbula amurensis*; Bax *et al.*, 2003).

Non-native species can be very common in urban regions. Proportions of non-natives vary from case to case, according to biogeographical region, ecosystem and scale under consideration, and type of organism, but in some cases they can dominate species assemblages. For example, Smetak *et al.* (2007) found no native earthworms at all in urban environments of Moscow, Idaho (USA), while 91 per cent of spontaneous vegetation on urban walls of Dunedin, New Zealand was non-native (de Neef *et al.*, 2008). Usually non-natives are less abundant, but observations in the order of 10–35 per cent non-natives are common (e.g. Loeb, 2006; Francis, 2011; Weeda, 2011)

Urban ecosystems therefore represent areas where species can establish and subsequently spread into surrounding regions. Successful establishment and spread will depend on many factors, including the number, frequency and scale of introductions ('introduction effort'), the time since introduction ('residence time'), whether predators or competitors are present in the invaded assemblage, and the particular characteristics of individual species, among other things (e.g. Richardson and Pyšek, 2006; Francis and Chadwick, 2012). The key aspects of invasion are rates of introduction and establishment, and the impacts that non-native and/or invasive species may have. There have been relatively few studies that have evaluated temporal changes in urban species assemblages over long timescales. In a rare exception, Tait *et al.* (2005)

examined changes in vertebrate and plant species in the urban region of Adelaide, Australia from 1836 to 2002. They found that there had been an overall increase in species richness of approximately 30 per cent, with 132 native species becoming extinct and at least 648 species colonising, most of which were plants. They recorded a decrease in native plant species by 7.5 per cent and an increase of 54 per cent due to non-native introductions, for an overall gain of 46 per cent. Mammals had an overall decline of 27 per cent (50 per cent decline and 23 per cent non-native introductions), while other vertebrate groups remained relatively constant (i.e. local extinctions were balanced by non-native introductions).

Most non-native species do not have significant environmental or economic impacts, and only a small proportion of species introduced become harmful. The 'Tens Rule' (Richardson and Pyšek, 2006) is often used to predict invasion success. Based on observations for plant species, it estimates that 10 per cent of species introduced to an area will establish in the wild; 10 per cent of those will become naturalised and will therefore be able to maintain populations indefinitely; and 10 per cent of those will prove to be harmful and thereby become 'invasive' or 'pest' species. Consequently, only 0.1 per cent of introduced species actually become harmful. The Tens Rule is only an estimate, however, and the range is usually somewhere between 5 and 20 per cent. Estimates of establishment rates are generally lacking for urban areas (though see Koch *et al.*, 2011), but may be higher than average due to high introduction effort, intentional cultivation and poor monitoring and control.

Impacts of invasive species include changes in biodiversity and assemblage composition due to: (1) direct (e.g. predation) and indirect (e.g. disease) mortality; (2) disruption of reproduction or resource use; (3) hybridisation or genetic introgression; and (4) environmental changes leading to a reduction in habitat provision and quality. Ecosystem services may also be compromised. Biodiversity impacts may be less significant in urban areas compared to more natural ecosystems, as they are atypical in their assemblages (see Section 6.4) and contain notable proportions of non-native species. Perhaps the most significant impacts in urban areas are economic and human health impacts (Pimentel *et al.*, 2005). A good example of the former is Japanese knotweed (*Fallopia japonica*), which has the capability to grow through tarmac and exploit minor cracks within road and building materials, and therefore compromise their structural integrity (Figure 6.1). Along with economic impacts related to repair and control, which can be substantial (a 2003 estimate of cost of removal of the species from the UK was £1.56 billion; Kabat *et al.*, 2006), the presence of the species has led to falls in house prices or even difficulty obtaining a mortgage for infested properties in the UK (RICS, 2011). Health impacts resulting from invasive species may be particularly significant in urban areas due to high population densities and potential exposure. Giant hogweed (*Heracleum mantegazzianum*), for example, is common in urban areas of Europe and is known for its phototoxic (light-reactive) sap, which causes skin inflammation and increased sensitivity to sunlight for several years (Pergl *et al.*, 2012). Invasive mosquitoes such as the Asian tiger mosquito (*Aedes albopictus*), again common in urban areas, may represent a major vector for diseases, including dengue and West Nile virus (Leisnham, 2012). The cumulative health impacts of such species can be substantial and are a key factor in increasing biosecurity measures in many countries.

A huge amount of effort goes into the management of invasive species. Most countries maintain committed screening and quarantine organisations, but the large

Figure 6.1 (a) Japanese knotweed (*Fallopia japonica*) growing through concrete at the base of a building in the United Kingdom, where the species is invasive. (Photo by Chris Cockel, used with permission.) (b) Japanese knotweed growing through scree in its native habitat and range (Japan, China and Taiwan). (Photograph by Rob Francis.)

number of species being transported both intentionally and unintentionally means that only so much can be done. Once a species is introduced and becomes established in the wider environment, it can be very difficult to eradicate. Full eradication is often only possible if a species is recognised as a threat, while its populations are low. Often impacts are only determined when a species has achieved a notable population, unless evidence of impacts in other host countries allows for the threat to be established early. In general, control options include: (1) physical control, including uprooting of plants, trapping and removal/killing of animals, shooting of animals, isolation of habitat by fencing, and so on; (2) chemical control, or the use of herbicides and pesticides to kill or disrupt activities such as reproduction; and (3) biological control, which is where a further non-native organism, usually from the host species' natural range, is released to control the invasive (Francis and Pyšek, 2012). In many cases combinations of treatments are utilised, and the intention is to limit population size and spread rather than eradication, which is often very hard to achieve. Each invasive species will require a specific suite of control activities to successfully limit population growth and further spread (Simberloff, 2008).

Urban ecosystems present particular conditions that further complicate control efforts. In many cases species are found on private land (e.g. gardens) and are actively cultivated, or perhaps not recognised as invasive – making coordinated control particularly problematic.

Even on public urban land there is often uncertainty as to the presence or extent of invasive species. Further, there are often problems identifying the responsible parties (e.g., who will perform the control, and who will pay for it?). Physical control can be difficult in densely populated areas, as can chemical control due to limitations or concerns regarding chemical use or lack of access to sites where the species can be found. Biological control requires extensive trials before release is sanctioned in many cases, and the heterogeneity of urban areas may sometimes limit the effectiveness of spread of the control agent. All of these factors may contribute to the abundance of non-native and invasive species in urban areas, though they remain an important focus for control efforts. In some cases, invasive species are supported by particular urban resources, and changing those resources or their availability to invasives may help with control (Box 6.1).

Humans are more than just a vector for the introduction and control of non-native species; the concept of non-native species is founded on an idea of naturalness that is socially constructed. Important questions remain to be answered regarding the nature and meaning of 'nativeness'. These include the emotional attachments to 'native' or 'traditional' biotic assemblages and the justification for vilifying successful non-native species as 'invaders' (possibly an instinctive xenophobic response). Clearly when considered objectively, successful invaders could be regarded as ecologically adaptable species that are being dispersed by anthropogenically mediated mechanisms and simply found beyond their usual range (Trudgill, 2008; Stromberg et al., 2009). Although the issue of economic damage can be relatively easily estimated for some species (Pimental et al., 2005), ecological impacts are harder to confirm, even for well-researched invasives (Stromberg et al., 2009). Questions also arise regarding societal views and treatment of urban non-natives, such as why some gain greater acceptance than others and how this may frame societal response to their perceived threat. A recent example of a non-native that has divided opinion within the UK is the monk

Box 6.1 Invasive crows in Singapore

House crows (*Corvus splendens*) (Figure 6.2) are an invasive species in many parts of the world, for example Singapore, where they have arrived after being introduced into Malaysia in the 1800s to control caterpillars in farms. There are many thousands of crows in Singapore, mainly in urban areas, and they are considered a pest as they can roost in groups of up to 3000, causing lots of noise and droppings, and can attack humans. They may also pose a threat to native bird species. In other parts of the world (e.g. Yemen) attempts to control the house crow by, e.g. poisoning them, have failed as populations are replenished by nearby sources. In Singapore, shooting has been the main method of control but this is ineffective and there is now an understanding that to reduce the populations, crows need to be deprived of their food and nesting sites. Soh *et al.* (2002) examined what urban environmental factors were most linked to house crow nesting sites throughout Singapore. They found that crows have a preference for nesting in yellow flame trees, as they have a large and dense crown with upward pointing V-shaped terminating branches useful for securely fastening the nests – the leaves also provide a balance between shelter and a good defensive view. However, this tree is also very common in Singapore – and house crows are very flexible in their behaviour, and so might easily use other trees (or human-made structures) if this species was not available. The most important factor involved in house crow nesting sites was urbanisation – crow nests were

Figure 6.2 The house crow (*Corvus splendens*), which is invasive in several parts of the world, including Singapore.

much more common within or adjacent to human habitation. This was probably linked to the availability of food – crows do not need to travel far to obtain
enough food to feed themselves and their young in places where humans live.
Heavy pedestrian and vehicle traffic not did deter crows from nesting, showing
further behavioural flexibility. The urban environment therefore provides a very
suitable habitat for house crows, and they have the ability to utilise this habitat
effectively. In order to prevent this species maintaining high abundances, Soh
et al. (2002) suggested that the urban environment would have to be manipu-
lated to make it less favourable, e.g. planting palm species which do not have a
similar morphology to the yellow flame tree, especially near rubbish dumping
sites, or regularly pruning trees to prevent nesting. Ensuring that refuse is properly stored in bins and therefore not accessible as a food resource was also
considered important.

parakeet (*Myiopsitta monachus*), which has spread from London into adjacent counties, with associated population increases, and since April 2010 can be culled by landowners under general licence 'to kill or take certain birds to preserve public health or
public safety' from Natural England. Some UK conservation organisations (e.g. the
London Wildlife Trust, Royal Society for the Protection of Birds) object to inclusion
in this list (misrepresented somewhat by the media as an extensive 'cull'), arguing that
the species is essentially an assimilated immigrant and that the decision is both xenophobic at heart and based on insufficient evidence (e.g. Barkham, 2009). Yet the species is a global urban invasive and pest species, against which a rapid response is
likely to reduce future economic and quality-of-life impacts for the London region
(Strubbe and Matthysen, 2009).

Non-native and invasive species present an ongoing debate for environmental
ethics, but as the development of an ecological understanding of non-native invasives
requires the role of humans to be incorporated into research, so too does the debate in
environmental ethics need to be supported by quantitative evidence of scale of the
problem, impact and response. 'Nativeness' is a somewhat questionable defence, especially in the urban context, where many species and assemblages are novel to the
urban environment. Urban non-natives are certainly here to stay, and this is increasingly being recognised, for example in the concept of 'recombinant assemblages'.

6.4 Recombinant assemblages

The combination of abundant generalists, limited specialists and the replacement of
native species with non-native species creates unusual and complex species assemblages in urban regions. These unique groupings have been termed 'recombinant'
assemblages (e.g. Soulé, 1990; Meurk, 2010). The concept of novel assemblages that
have emerged following anthropogenic modifications of ecosystems, and which therefore have no 'typical' classification or habitat associations, is also echoed in the 'novel
ecosystems' of Hobbs *et al.* (2009), or 'emerging ecosystems' of Milton (2003). These
concepts represent a growing recognition that atypical assemblages are becoming the

norm with both changing environments and the spread of non-natives, and as a result societal attitudes to these species may need to be revised (Davis *et al.*, 2011). This is particularly relevant to urban areas – there is certainly an argument to be made that the specific species found within an assemblage are less important than the types of species found and whether the ecosystem functions well.

Meurk (2010) classifies four main types of recombinant assemblage combinations, though also notes that as recombinant communities are in reality 'infinitely complex', the combinations are not necessarily discrete and unchanging. The first three are:

1   remnant, in which the soil is not artificial and remains unimproved (i.e. has not been cultivated) and wherein there is some similarity between the 'original' community and the new assemblage, for example due to a long-standing seed bank;
2   spontaneous, in which artificial substrates are colonised by species often found in disturbed environments;
3   deliberative, in which species are purposely planted (or sometimes removed) to create a desired assemblage, e.g. for landscaping, gardens or restoration projects.

The fourth classification is a 'complex' assemblage, which incorporates elements of two or more of the above. Clearly, the built environment is most likely to present 'spontaneous' assemblages, with engineered structures such as living roofs and walls also exhibiting 'deliberative' assemblages. Most gardens, parks and infrastructure may be predominantly 'deliberative' but will also contain elements of the other classifications, while brownfield sites will be mainly spontaneous but with perhaps some remnant species from the seed bank. Aquatic ecosystems are often not considered, but are primarily combinations of spontaneous and remnant assemblages, unless species are specifically planted or introduced as part of a restoration scheme.

6.5 Urban succession

Many ecosystems undergo a process of change termed 'succession', as it involves the replacement of one species assemblage with another in a repeated progression over time (see Figure 6.3). A 'sere' is a full cycle of development, and each individual step of assemblage change in the progression is termed a 'seral stage'. Much work has been conducted into seral progression since the early twentieth century, and several hypotheses have emerged in relation to the key factors driving succession, as well as whether the idea of a stable 'end' or 'climax' assemblage is valid or not (for a review see McCook, 1994). Concepts have changed from assuming relative stability and predictability of process and climax state, to a recognition of inherent variability and unpredictability in many elements of succession (Francis, 2009b). This may be particularly true in urban ecosystems due to the recombinant assemblages described above, which further complicate the stages and processes observed. Despite this, key trends may be generalised, and considered in the context of urban ecosystems.

6.5.1 Disturbance and stress

Disturbance in urban ecosystems usually involves the destruction of biomass, such as the physical removal of an area of vegetation or soil, whether via natural (e.g. fire,

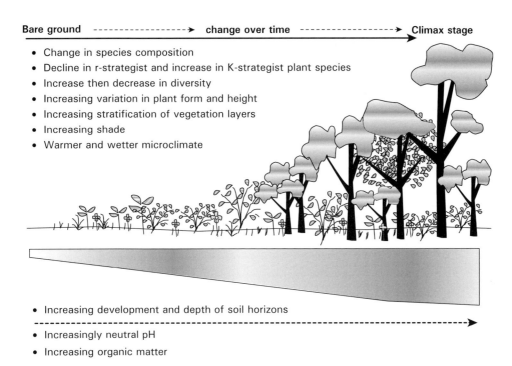

Bare ground ----------------→ change over time -----------------→ Climax stage

- Change in species composition
- Decline in r-strategist and increase in K-strategist plant species
- Increase then decrease in diversity
- Increasing variation in plant form and height
- Increasing stratification of vegetation layers
- Increasing shade
- Warmer and wetter microclimate

- Increasing development and depth of soil horizons
- Increasingly neutral pH
- Increasing organic matter

Figure 6.3 Ecological changes in an ecosystem over time resulting from vegetation succession.

landslide) or anthropogenic (woodland clearance, construction) processes. Such disturbances are important for allowing succession to occur, as it creates the bare 'starting ground' for colonisation and assemblage development. Some disturbances are very common in urban areas, such as both the construction and destruction/abandonment of the built environment, which creates frequent disturbances and sites for colonisation. Urban management practices also frequently disturb assemblages (e.g. the mowing of roadside verges) that may influence varying stages of succession. Contrastingly, some disturbances that are detrimental to humans, such as fires, flooding or mass movements of soil, are reduced in urban environments through active management (e.g. Zipperer, 2010). Succession is usually classified as either primary or secondary, with primary succession taking place on a substrate that has not supported vegetation before, and is therefore undeveloped and poor in fine sediment and organic materials, with no seed bank or soil organisms. Secondary succession takes place on an area that has previously supported organisms and has since been disturbed. The soils and seed banks are therefore more developed and the early stages of secondary succession take less time. Both forms are found in urban environments. Although primary succession is usually associated with newly formed substrates from glaciers and volcanoes, many urban soils are composed of artificial materials that have never before supported organisms and are therefore similar to primary substrates (see Section 3.7; Pouyat *et al.*, 2010; Zipperer, 2010). Secondary succession will more usually occur in green spaces following disturbance.

Stress, on the other hand, is a factor that limits the growth or performance of an organism, such as a surfeit or deficit of water, light, heat, nutrients and so on. Urban environments are stressed by elevated temperatures, chemical pollutants, and often either too much or too little soil water (see Chapter 3), with individual sites having particular problems. These stressors will limit the species that can colonise sites and thereby partly determine the types of assemblages that may form.

6.5.2 Colonisation

Colonisation occurs rapidly in urban environments due to the abundance of organisms as discussed above. However, it is likely that most colonisers will be generalists and opportunists (e.g. ruderal plant species), with the specialists found drawn from a smaller regional pool than in a more natural ecosystem. There are many vectors for colonisation and although there are limited studies of species dispersal, it is clear that many individual seeds and organisms are continuously moving around the urban region. Although humans will have a greater role in species dispersal here than elsewhere (see Zipperer, 2010), the majority of plant colonisers arrive via wind or attached to animals.

6.5.3 Environmental modification

The growth of plants and the utilisation of a site by animals and microorganisms will inevitably result in modifications to that environment. Such modifications may lead to the 'facilitation' of other species and assemblages that are more suited to the modified environment, and are consequently associated with species replacement as described below. For example, the breaking up of soils and the addition of organic matter from dead plants or animals will help to increase moisture and nutrient availability, favouring competitive species over those that may be able to cope better with stressful conditions or frequent disturbance. As a rule, the general trends in modification of the physical environment are associated with increasing vegetation cover, and include the breakdown of rocks and soil formation, increasing soil organic matter, movement to a more neutral soil pH, reduced light availability, and increasing temperature and relative humidity. These modifications are all likely to occur in urban ecosystems, unless continued disturbance prevents succession from occurring at all.

6.5.4 Replacement

Replacement of a species can occur via environmental modification or by the more stochastic arrival of species that may outcompete or otherwise replace species that are currently present. Some replacement may occur from species that have been present in the ecosystem from the start but have small populations or (in the case of some plants) are slow-growing and simply become more dominant over time. It is debatable what the key replacement processes are, and these are likely to vary from site to site (Meiner and Pickett, 2011).

General patterns emerge due to replacement. First, increasing variation in plant growth form along with increasing vegetation height, which together result in

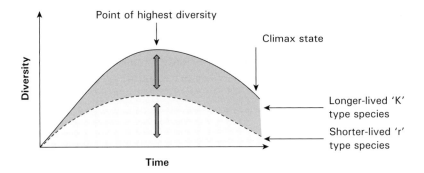

Figure 6.4 Change in species richness over time as succession takes place, highlighting that richness is highest in the middle stages. The relative proportion of 'r-strategist' and 'K-strategist' species changes over time with r-strategists dominating the assemblage in the early stages of succession and K-strategists at the later stages.

stratification of vegetation and a more heterogeneous structure. Second, an abundance of opportunist 'r-strategist' plants in the early stages with a gradual replacement by slower-growing and more competitive 'K-strategist' plants. Lastly, an increase and then a decrease in species diversity, such that the highest diversity is found in the mid-stages of succession (Figure 6.4). Overall, urban replacement trends are hard to predict. The abundance of generalists and opportunists will mean that the initial stages of urban succession are likely to follow typical trends, though assemblages may be drawn from a wide range of species due to the large number of such species in the urban environment. Further stages and trajectories of change are less predictable and will vary depending on ecosystem type and condition (see Chapter 4).

Understanding urban succession may be important, for example, when trying to ensure available habitat for species of conservation concern, maintaining ecosystem services, and for urban spatial planning. Kattwinkel *et al.*'s (2011) concept of 'temporary conservation', for example, explicitly argues that a regional species pool may be supported by a range (mosaic) of different seral stages within the region.

6.6 Synurbic species

Some species that particularly associate with areas of human habitation, and yet are not domesticated, are often termed 'synanthropes'. Such species enjoy extensive ranges or dramatically increased populations due to their associations with humans, and often may become weeds or pests. Many of these species are populous on farmland or other characteristically anthropogenic ecosystems, and some species become preferentially associated with specifically urban areas. Francis and Chadwick (2012) label the latter category of species 'synurbic' (after Luniak, 2004), and suggest that this term should be applied to those species, or rather populations of species, that maintain denser populations within urban environments compared to their native environments, and therefore demonstrate a quantifiably preferential association with urban regions.

A range of species are likely to be synurbic, though this remains to be determined in many cases – see Francis and Chadwick (2012) for some examples. Many animals specifically associated with humans are experiencing urban population booms – a good example is the various species of dust mite (family: Pyroglyphidae) found in our homes. These mites live on small organic particulates and are consequently found in house dust, and in stored dried food products. They are a source of allergens associated with diseases such as asthma, rhinitis and dermatitis. These diseases are also on the rise in urban environments – this may be linked to various causes, but a likely one is high dust mite populations and densities. Research into dust mite habitat in urban environments, both indoor and outdoor, has shown that the natural habitat of mites is bird nests, but human habitations are suitable analogues for this habitat, and easily invaded. Indoor samples from houses in Poland found mites in just over half of the samples taken (56 per cent) with the Pyroglyphidae (allergenic house dust mite) family being dominant (96 per cent of mites found) (Solarz *et al.*, 2007). Other likely synurbics of increasing concern are bed bugs (see Box 6.2).

Box 6.2 The resurgence of bed bugs as an urban pest

The bed bug represents a synurbic pest that is dramatically on the increase, and its resurgence has been described as a 'perfect storm' for pest control (Potter, 2005). There are two main species of bed bug that associate with humans: *Cimex lectularius* (a temperate species with pan-global distribution; Figure 6.5) and *Cimex hemipterus* (a tropical species, again with pan-global distribution). The two do overlap in range and can mate, but do not produce viable offspring – this is given as a reason for *C. lectularius* being largely absent from parts of East Asia, where interspecific mating does sometimes occur. This would suggest that there were two separate populations some time ago but that the movement of people around the globe has expanded their range so that they have now become reacquainted. They are cryptic species, and establishing speciation origin has not yet been possible – they have been synanthropes and feeding on humans since our species first utilised cave systems. *Cimex lectularius* may actually be an amalgamation of subspecies rather than a distinct species by itself.

Cimex lectularius in particular was a major pest species until after World War 2 and the development of (e.g.) DDT and more advanced cleaning technology in more developed regions; as a result, they have effectively passed out of 'living' memory (Reinhardt *et al.*, 2008). From the 1950s to 1990s they were mainly prevalent in less developed regions, though occasional outbreaks were reported in developed countries (Boase, 2008). Probably their re-emergence (1990s onwards) in developed countries is due to: (1) increased rates of human immigration and tourism, as people often act as vectors for the species; (2) the decline in home pesticide use combined with increased pesticide tolerance due to natural selection of remnant populations; and (3) the lack of public knowledge of the species and techniques to prevent spread and manage populations. It is thought that 'inner-city' or poor residential urban areas may have acted as a reservoir for the species and been the main source of re-emergence (Boase, 2008). Health impacts of the species can be significant – although they do not

Figure 6.5 The common bed bug (*Cimex lectularius*), which is experiencing a global resurgence. The average bed bug is around the size of an apple seed. (Photograph by James Logan, used with permission.)

act as disease vectors, they are associated with dermatological effects, allergies, unpleasant odours due to oily chemical secretions, and in particular have negative psychological effects. This makes them particularly significant for tourism and recreation industries, as they are not 'harmful' enough to be immediately actionable (like a disease outbreak), but are likely to negatively affect businesses, particularly given that social tolerance of the pest is very low. The species tends to rest in dry materials and prefers wood and paper over stone, plaster, metal or textiles. They are often found in beds and bedding, but also in crevices in furniture, clothing, carpets, curtains and wall-hangings, and are not just found in homes and bedrooms – a range of public locations may harbour the species, including public transportation (Potter *et al.*, 2010). Full eradication is difficult, particularly when use of pesticides may be restricted, but is certainly possible – though repeated treatment may be required (Boase, 2008). Given the high rate of emergent cases (Boase (2008), for example, notes an increase of almost 25 per cent per year in reported cases in London from 2000 to 2005), and the difficulties in treatment and prevention of spread due to a reduced cultural awareness of the pest, bed bugs are likely to be a dominant urban pest for the foreseeable future.

6.7 Urban pests

Some of the most populous and widely recognised urban species (including some synurbics) have notoriety as pests. A pest is usually defined in terms of economic damage to specific crops or other form of natural resource (e.g. Sileshi *et al.*, 2008), but in an urban area the definition may be broadened to include impacts that lower quality of life, including health impacts, economic damage to property, excessive noise, emotive repulsion and so on. In many cases, the close association of certain species with humans, making them synanthropic or synurbic, inevitably leads to them becoming a problem where human populations are concentrated. Common pests include various groups of organisms, such as species of ant (family: Formicidae), bird (class: Aves), fly (order: Diptera), bees and wasps (order: Hymenoptera), rodent (order: Rodentia), spider (order: Araneae), mosquito (family: Culicidae), termite and cockroach (order: Blattodea) and bed bug (genus: Cimex). Some common urban pests are often non-native to a given region, for example the widespread brown rat (*Rattus rattus*) and the Asian tiger mosquito (*Aedes albopictus*). Pests are consequently some of the most well-researched urban species, though much focus has been on control and extermination. Despite increasing efforts on finding ways in which people and pests may live together more easily, given the limitations of many forms of pest control, the focus still remains on ways in which pest populations and distributions may be limited.

The distribution and abundance of pest species is often linked to socio-economic parameters (e.g. Masi *et al.*, 2010; Unlu *et al.*, 2011). Abundances of invertebrate pests such as mosquitoes and cockroaches have been consistently (though of course not exclusively) linked to poorer locations in urban regions (Bradman *et al.*, 2005; Unlu *et al.*, 2011), as have other pests such as rodents (e.g. Masi *et al.*, 2010). In many cases this is due to age of housing and infrastructure creating suitable niches and conditions for species to persist, inability to pay for or implement pest control or prevention, and increased population density and household size. Particular urban areas may therefore maintain some of the largest populations of problematic urban species, and act as source locations for spread and establishment of the species in other parts of the complex (see Box 6.2).

6.8 Species responses to urban environments

There are several ways in which species may respond to urban environments, some of which may lead to synurbic populations (this process being termed 'synurbisation'; Francis and Chadwick, 2012). These include behavioural, phenological, morphological or physiological responses, though individual species often display more than one type of response, or the responses overlap. For example, the modification of the house finch's (*Carpodacus mexicanus*) bill to feed on the seeds of plants abundant in urban areas (Badyaev *et al.*, 2008) is both a morphological and a behavioural response, and a further behavioural response may be seen in the species' variation in courtship songs between urban and rural populations. Some examples of the types of responses demonstrated by urban species are given in Table 6.1, and an example of behavioural responses of urban foxes is given in Box 6.3.

One important question is whether such responses may have a genetic basis or not, as this may have implications for processes of natural selection and the future of

Table 6.1 Examples of observed responses to urban environments by members of different organism groups

Type of response	Cause of response	Organism groups
Morphological change, e.g. beaks or plumage/colouring	Altered selection pressures	Birds, invertebrates
Phenological change, for example flowering, seed set, foraging time or duration, earlier and/or prolonged reproductive season	Urban climate and microclimate effects e.g. increased temperatures, longer seasons	Birds, mammals, invertebrates, plants
Reduced migration	Abundant resources and longer seasons reduce migration cues or necessity	Birds
Changes in frequency and pitch of birdsong	Birds sing differently to be heard above background urban noise	Birds
Use of anthropogenic structures and/or food resources	Abundance of the built environment and uneaten food waste (for example) creates opportunities for species to find alternative habitat, e.g. nesting sites in buildings	Birds, mammals, invertebrates, plants
Increased human tolerance and reduction of escape distances	Proximity to humans means that species come to regard them as less of a threat than less-exposed organisms in non-urban environments	Birds, mammals
Reduced territories and foraging area/duration	Abundant resources (e.g. uneaten food waste, available mates) mean that territories are less important and animals may not need to forage for as long as in more natural ecosystems.	Mammals
Communal or pack living	Reduced resource competition allows animals to live more closely than they might in their natural ecosystems	Mammals

Summarised from Francis and Chadwick (2012) and references therein.

urban species. A response that is associated with genetic change may be considered an 'adaptive' response (Francis and Chadwick, 2012). Most successful urban species, including those that may be considered synurbic, seem to respond to their environment within their natural range of plasticity (Bokony *et al.*, 2010; Evans *et al.*, 2011). This means that they have sufficient flexibility of physiology, behaviour and so on to exploit a range of environments without necessarily undergoing any process of natural selection for useful characteristics (adaptation). Many of the responses listed in Table 6.1 are within the boundaries of plasticity exhibited by many species in non-urban environments. For example, territory or range sizes are very flexible in some species, and so decreases in these reported for some urban mammals (e.g. European fox (*Vulpes vulpes*) or European badger (*Meles meles*)), may be considered a response to an urban environment, but not necessarily a form of adaptation. Likewise, changes to birdsong may mean that the birds in question are simply using a different selection of their vocal portfolio, rather than evolving any new capabilities to cope with the

---

Box 6.3 Urban foxes

The European fox (*Vulpes vulpes*) is a ubiquitous urban species within Europe, and a good candidate to be synurbic (Francis and Chadwick, 2012). People's opinions of them vary widely – for some they are an unwelcome pest and potential disease vector (e.g. rabies), while for others they are an interesting and lovable urban companion. They are one of the typical 'urban' species that has been studied to a wide extent, particularly the ways in which they have responded to urban environments. Not only have urban foxes exhibited high population densities in urban regions (with key studies in Zürich, Switzerland and Bristol, UK: Baker *et al.*, 2000; Gloor *et al.*, 2001), but reduced territory size, increased tolerance of humans, foraging in packs, and the ability to utilise human food sources have also been observed. This latter factor may be responsible for all the other observations. Contesse *et al.* (2004) examined the stomach contents of 402 foxes shot or found dead in Zürich during 1996–98, and separated contents into natural (e.g. rodents, birds, other vertebrates, invertebrates), anthropogenic (e.g. scavenged meat and waste food, pets and pet food) and intermediate (e.g. cultivated fruit and crops) food sources. They found that the majority of foxes were utilising anthropogenic foods, this being on average 53.6 per cent of stomach contents. A further 18.2 per cent of average stomach contents was cultivated fruit and crops. Very few stomachs contained a single food source, the most common (9 stomachs) single food source being scavenged meat – 7 stomachs contained only wild rodents and 3 only invertebrates, which are natural fox food. This surfeit of available food means that a large fox population can be maintained, that the need to hunt alone to retain stealth is lost, that individual hunting ranges are no longer relevant so that territories are less important and scavenging in packs is more common, and that foxes are more prepared to get close to humans to access these food sources. This is a good example of how a species can alter its behaviour to take advantage of a plentiful urban resource.

---

urban environment. Indeed, Møller (2009) and Hu and Cardoso (2010) suggest that it is actually those species with high plasticity that can most easily cope with the challenges of urban environments, and are therefore 'preselected' for urban success. Of course, it is possible that the expression of certain traits within plasticity may lead to selection over time, and this is one of the problems with establishing whether urban species are adapting or not. In theory, genetic differentiation between urban and non-urban populations of a species should suggest this, but often differentiation can be found due to geographical isolation of urban populations occurring from urbanisation (e.g. the isolation of remnant populations as the built environment is constructed) or from founder effects (where the traits of the few individuals that first colonise an area become prevalent in the future population). Potential founder effects have been found for urban blackbirds (*Terdus merula*) and European foxes (*Vulpes vulpes*) (e.g. Wandeler *et al.*, 2003; Evans *et al.*, 2009). Byrne and Nichols (1999) note that the

London underground mosquito (*Culex pipiens* f. *molestus*) is genetically distinct from aboveground *Culex pipiens* populations, and the limited genetic diversity of the underground populations suggests that colonisation of the underground network may have been a single or at least very rare event. This further blurs the distinction between plastic and adaptive responses.

Some studies have suggested adaptive responses, however, and 'microevolution', or small genetic changes over a short period of time that leads to the emergence of a new form, variety or subspecies, has been observed. The London underground mosquito may be an example of this, particularly due to its prevalence in underground ecosystems in varying urban regions round the world that might suggest selection rather than dispersal. The changes in bill morphology and seed consumption that have been observed in urban populations of *Caropdacus mexicanus* are supported by genetic differentiation and have been attributed to microevolution (Badyaev *et al.*, 2008), though this has not necessarily led to the emergence of a new form or biotype.

Species that respond positively to urban environments, and which may or may not become synurbic, are important for urban ecosystems because they are likely to feature prominently in urban assemblages and are the most likely candidates for the emergence of new forms or species. As wider regional and global biodiversity is lost, and more and more species are brought into urban areas that may be able to adapt and interbreed, questions arise as to whether urban regions may form the nuclei for expansion of genetic variability and future speciation.

6.9 Chapter summary

- Urban species assemblages are broadly characterised by increases in proportions of generalist species and decreases in specialist species, with some exceptions such as those specialists associated with rock or cliff habitats.
- Many urban species colonise from outside the urban complex, and a significant proportion (usually 5–35 per cent) of urban species are non-native. Urban regions therefore represent locations where non-natives, and harmful invasive species, may establish and spread.
- As a result, many urban species assemblages are non-typical and may be termed 'recombinant' assemblages. There are different types of recombinant assemblages, depending on whether species have been intentionally planted or introduced, or have emerged spontaneously.
- All ecosystems and assemblages change over time, and succession is one process via which this occurs. Succession in urban ecosystems may be particularly complex due to recombinant assemblages and alterations to disturbance, stress and resource abundance.
- Synurbic species are those that maintain denser populations within urban environments compared to their native environments. These may often include urban pests.
- Species may respond to the urban environment via changes in behaviour, phenology, morphology and/or physiology. It remains to be determined whether such changes may lead to adaptation and the emergence of new urban biotypes or species.

6.10 Discussion questions

1   List the species you most associate with urban environments. How do they fit into the varying categories described here (generalist, specialist, non-native, synurbic)?

2   Are any invasive species a particular problem in your urban region? Which ones, and why? Would you recognise the species if you saw it? How is it being controlled?

3   Are 'recombinant assemblages' and 'novel ecosystems' a good or bad thing for global biodiversity, and why?

4   Have you experienced contact with any urban pests? How important do you feel pest control is for maintaining a healthy urban environment? Are there any positive impacts of pest species on the environment?

5   What are the implications of genetic divergence and natural speciation within urban ecosystems? To what extent could urban ecosystems be considered 'havens of biodiversity'?

Chapter 7

Nature conservation in urban regions

7.1 Introduction

The notion of nature conservation may seem inappropriate for urban ecosystems upon first consideration. As urban environments are often complex mosaics of different assemblages, often with limited 'naturalness', the idea of preserving or even improving such areas may seem either nonsensical or a waste of effort and resources. Chapters 3–6 have demonstrated that urban ecosystems may have unexpectedly high diversities, and that the species and communities within such systems may be interesting, unusual and ecologically valuable. This is a relatively recent viewpoint, however, and the history of conservation has for the most part ignored urban regions; perhaps because 'urban' spaces were characterised by a perceived absence of 'nature' (Williams, 1973). This has begun to change in recent years, as both researchers and the wider public have become more aware of the value of urban nature and ecology, and also the potential for urban nature conservation to allow society to respond to environmental change (e.g. climate change), support environmental education, ecosystem services, urban sustainability, and human health, and to behave ethically in relation to the environment. As noted in Chapter 1, there can be substantial resources brought to bear on urban conservation, and there is great potential for both ecological and societal improvement of urban ecosystems.

There are three broad forms of conservation, which have varying relevance for urban ecosystems:

1   The *preservation* of genotypes, species and ecosystems (including the distinct communities associated with certain ecosystems). This is achieved by the isolation and shelter of organisms directly in zoos and botanical gardens (for genotypes and species), the taking and storage of genetic material for future reference, e.g. seed vaults (for genotypes and species) or the establishment of reserve areas that are protected from future harm (for genotypes, species and ecosystems). Preservation is mainly applied to natural or semi-natural areas of the globe and has relatively limited application to urban areas, though urban green spaces are sometimes preserved in recognition of their ecological or societal value (see Chapter 4).

2   The *restoration* of ecosystems and reintroduction of extirpated species. Restoration is achieved by either removing factors that have caused degradation or by the full re-creation of appropriate physical conditions and installation of

communities that have previously been lost. Urban restoration and its varying permutations (rehabilitation, etc., discussed below) is quite common, though tends to have greater constraints than in non-urban areas. Species reintroductions are relatively rare in an urban context, but may occur in tandem with ecosystem restoration.

3   The ecological engineering of anthropogenic ecosystems to encourage both societal use and biodiversity. This is often termed *reconciliation ecology* (see Rosenzweig, 2003; Francis and Lorimer, 2011), and has great potential in urban areas. The concept of reconciliation ecology incorporates the view that although ecological preservation and restoration are highly valuable, they are insufficient to address the massive scale of habitat loss and degradation that the natural world has experienced due to anthropogenic activities, and that therefore ways of reconciling human and non-human land use must be found. Urban areas are prime locations for this; they have limited capacity for preservation and restoration, but ecological and environmental engineers are continually finding ways to incorporate biodiversity and its attendant benefits into urban design (e.g. Chapman and Underwood, 2011; Francis and Lorimer, 2011). 'Ecological engineering' refers to the creation of ecosystems or habitat, and is distinct from 'environmental engineering', which is usually the improvement of environmental quality via control of anthropogenic wastes and pollution. In some cases engineering techniques may do both: the examples of living roofs and walls have been covered in Chapter 5.

Terms relating to these forms of conservation are summarised in Table 7.1. Note that there is often conflation between the different terms and the ways in which they are applied. For example, deer control in an urban woodland reserve may be considered either 'reserve management' or 'restoration', depending on the context and the particular objectives set. Likewise, the placing of dead wood within an urban park to create habitat may be considered management, restoration or reconciliation. The three aspects of conservation are now considered in more detail with specific reference to urban ecosystems.

7.2 Ecological preservation

Green spaces are most likely to be selected for preservation, though in many cases they are not formally designated as nature reserves, and have rather been preserved fortuitously. The designation of particular urban ecosystems as ecological reserves is not unusual, though this has historically focused on large areas of parkland or remnant woodland rather than smaller areas of green space (Ranta *et al.*, 1999; Ooi, 2011). Most large urban regions contain at least some green space that is officially designated as one or more protected areas (e.g. George and Crooks, 2006). As noted in previous chapters, larger reserves are generally found to have the highest ecological value due to species–area relationships and the presence of 'interior' habitat that is more likely to support species missing elsewhere in the urban complex (e.g. Godefroid and Koedam, 2003). They may be particularly significant for urban species populations, as reserves are more likely to maintain abundant resources and therefore act as 'source' ecosystems within the urban region. In some cases, large reserves within or

Table 7.1 General definitions of restoration ecology terms

Terminology	Definition
Restoration	Returning an ecosystem to a former pre-disturbance state and/or to an unimpaired condition
Rehabilitation	Partial return of an ecosystem from a degraded state towards a pre-disturbance state, or a different but sustainable condition
Reconciliation	Modification or improvement of anthropogenic ecosystems to encourage both societal use and biodiversity
Remediation	Removal of a detrimental influence (e.g. pollution) to allow an ecosystem to function better or maintain a higher ecological quality
Replacement	The creation of an alternative (new or heavily modified) ecosystem or ecological state
Enhancement	An improvement in environmental condition or ecological quality without consideration of a previous state
Mitigation	Moderation or lessening of an impact on environmental or ecological condition or quality

Based on Hamby (1996), Bradshaw (1997), Rosenzweig (2003) and Francis (2009a).

around urban complexes may prevent further urbanisation (see Chapters 2 and 8) and have particular value with regard to this 'safeguarding' role.

Urban reserves will support higher biodiversity and ecological quality if human activity is minimised. Even relatively light recreational use can create sufficient disturbance to impact upon the use of the habitat by sensitive species, such as some avifauna or mammals (e.g. George and Crooks, 2006). However, many significant societal benefits are also associated with large areas of green space, which are often designated as reserves. This can create conflict over green space use in urban areas, with a fine balance sometimes to be drawn between human access and habitat provision. The absence or limitation of human use also makes provision of habitat variety and preservation of structural elements such as standing dead wood more feasible, both of which have been shown to be important for biodiversity (see Chapter 2).

More recently, there has been greater recognition of the conservation value of other forms of green space, and gardens and brownfield sites have been incorporated into conservation policy and practice (Gaston *et al.*, 2007; Goddard *et al.*, 2010). In some cases built ecosystems may be considered worthy of conservation for their ecological value, such as the tree-supporting masonry walls of Hong Kong (Jim and Chen, 2010).

The establishment of conservation areas is determined by local, regional or national government, and is primarily based around key legislative instruments such as the Convention on Biological Diversity (CBD) and its attendant national biodiversity strategies and action plans (see www.cbd.int), as well as other legislative or policy instruments (see Table 7.2). Despite often being set at a national level, much actual management is determined by local government, regional branches of environmental non-governmental organisations (ENGOs), and the all-important local citizenry, who in many cases do much of the actual monitoring, maintenance and stewardship work on a voluntary basis (Krasny and Tidball, 2012). Much urban planning is based

Table 7.2 Key legislative and policy instruments used to inform and enact conservation activities around the world. Some brief examples are given of how these may relate to urban ecosystems

Legislative or policy instrument	Purpose	Link
Convention Concerning the Protection of the World Cultural and Natural Heritage (World Heritage Convention) 1972	*To identify and conserve areas of particular cultural or environmental significance, as part of humanity's shared heritage* Many historic urban areas feature in the list of World Heritage Sites, mainly for cultural reasons (118 sites are classified under 'urban')	whc.unesco.org
Convention on International Trade in Endangered Species of Wild Fauna and Flora (CITES or the Washington Convention) 1975	*To ensure that trade in wildlife specimens (from wild animals to animal products) does not threaten their existence* Trade may often be centred on urban regions	www.cites.org
The Convention on Wetlands of International Importance, especially as Waterfowl Habitat (the Ramsar Convention) 1975, 1982, 1987	*To maintain ecological quality and ensure sustainable use of wetland ecosystems* Historically, urbanisation has been particularly detrimental to wetlands, though wetland restoration and constructed wetlands are becoming more common	www.ramsar.org/cda/en/ramsar-documents-texts-convention-on/main/ramsar/1-31-38%5E20671_4000_0__
Birds Directive 1979, 2009 [79/409/EEC]	*To protect habitat for endangered and migratory birds throughout the EU, mainly through the designation of Special Protection Areas (SPAs), which together form a proportion of the NATURA 2000 network. Also to prevent certain activities that threaten bird species, including killing, nest destruction and trade* Urbanisation has been responsible for the fragmentation of much habitat for bird species, while some urban ecosystems may provide important habitat for particular species	ec.europa.eu/environment/nature/legislation/birdsdirective/index_en.htm
Convention on the Conservation of European Wildlife and Natural Habitats (Bern Convention) 1982	*To ensure that appropriate steps are taken to conserve wild (i.e. not domesticated) species and their habitat within the EU*	conventions.coe.int/treaty/Commun/QueVoulezVous.asp?CL=ENG&NT=104
Convention on the Conservation of Migratory Species of Wild Animals (Bonn Convention) 1983	*To conserve migratory species throughout their range, for example by removing barriers to migration and preserving or restoring species habitat. Agreements are made for*	www.cms.int/about/intro.htm

(Continued)

Table 7.2 (Continued)

Legislative or policy instrument	Purpose	Link
	particular species or species groups between countries that encompass their range	
	Urban areas contain some important habitat for migratory species, especially if urbanisation has fragmented a biodiverse area. They also represent key locations for habitat restoration to improve migration capability. Urban ecosystems may also reduce migration in some species by reducing the need to migrate due to resource abundance (Chapter 6)	
Directive on the Conservation of Natural Habitats and of Wild Fauna and Flora 1992 [92/43/EEC] (EU Habitats Directive)	*To conserve natural habitats of wild species to maintain sustainable populations. Key habitats are those that are at risk of disappearance, have a small natural range and are therefore more vulnerable to impacts, or are outstanding examples of key biogeographic regions*	eur-lex.europa.eu/LexUriServ/ LexUriServ.do? uri=CONSLEG: 1992L0043:20070101:EN:HTML
Convention on Biological Diversity 1993	*To ensure sustainable use of biodiversity, at genetic, species and ecosystem levels. This includes promotion of sustainable development, conservation measures, ecosystem rehabilitation and restoration, species recovery and reintroduction, and control of invasive alien species, among other things*	www.cbd.int
	Urban ecosystems may represent important species habitat, and are also key locations for ecological improvements and the control of invasive alien species	
Also note: IUCN and European Red Lists 1963–2012	*To identify species or groups of species under threat or at risk of significant decline or extinction, to inform conservation action*	www.iucnredlist.org/initiatives/ europe
	As the most comprehensive global database, this is used to inform action in most of the legislation and policy instruments noted above	

Note
All interpretations are ours – for more information on the specifics of each instrument, see links.

around the establishment of reserves (in combination with ecological restoration in some cases) for the preservation of endangered species, overall biodiversity and enhancement of ecosystem services (see Chapter 8).

Reserve establishment is most common in urbanising areas, though much depends on pre-urban land use and the presence of existing reserves, while restoration is more common in long-established urban areas. Often, both are performed to preserve specialist or more vulnerable species in response to legally protected endangered species lists and/or declining urban populations, rather than more common generalist species that may be found in urban regions. This may be particularly important for keystone species (Mills *et al.*, 1993), which are those that have a significant and important influence on their ecosystem or assemblage that is disproportionate to their abundance. Such species may play important roles in (for example) dispersing seeds and providing niches for other species. Consequently, providing sufficient habitat for keystone species may indirectly support populations of a wide range of species. Hougner *et al.* (2006) describe oaks (*Quercus robur* and *Quercus petrea*) in the Stockholm National Urban Park as keystone species, as they provide unique niches for up to 1500 other species and many endangered insects. They determined that the majority of acorns are dispersed (buried) by Eurasian jays (*Garrulus glandarius*), which act as a 'mobile link' or organism that can cross landscape boundaries to exchange seeds (among other things) between different ecosystems. Conserving Eurasian Jay populations is therefore crucial for maintaining populations of the keystone oaks in this urban reserve, with the cost of replacement seeding/planting by humans in the absence of jays estimated at between approximately US$ 21,000 and US$ 94,000 per hectare.

Where reserve creation is not possible or is limited, restoration and reconciliation efforts become more important.

7.3 Ecological restoration

Ecological restoration essentially involves returning a degraded or modified ecosystem to a former (and presumably more desirable) state. Ideally, restoration should return a system to its state of existence before anthropogenic disturbance. Such complete restoration of an ecosystem is often difficult to achieve, and not always appropriate; consequently, most ecological 'restoration' is actually rehabilitation (see Table 7.1). Both of these terms essentially involve some form of ecological improvement, which may be evaluated based either on social (i.e. what people want, or value) or functional (i.e. biodiversity, rates of biogeochemical cycling, ecosystem services) grounds. The key difference between restoration and rehabilitation on the one hand, compared to reconciliation, enhancement or replacement on the other (see Section 7.4), is that the former generally involve an improvement that brings the degraded system closer to its pre-anthropogenic state. The latter may involve societal or ecological improvement, but do not have a previous state to act as a 'guiding template' for setting objectives or evaluating success. Mitigation or remediation usually involves the removal of a factor that is limiting diversity, ecosystem function or societal use (such as pollution) and may help to rehabilitate a system, but does not necessarily involve the return to a former state. For example, the construction of the London Olympic Park (see Box 7.1) involved much remediation of degraded soils, but did not involve a return to a pre-industrial state. Rather, the initiative led to the creation of a new series of urban

Box 7.1 The London Olympic Park development

The hosting of the 2012 Olympic Games in London necessitated the construction of games venues at a site in east London, termed the 'Olympic Park'. The site was located in the downstream portion of the River Lee catchment, which is a tributary of the River Thames, along which runs a greenway termed the Lee Valley Regional Park. Development of the Olympic Park resulted in a loss of 45 ha of existing conservation areas, mainly brownfield sites, which was to be mitigated by: (1) protecting some retained habitats; and (2) creation of new and improved habitat within the site. A Biodiversity Action Plan (BAP) for habitat preservation and creation was approved in 2007, and included habitat found prior to construction (built environment, 0.40 ha; allotments, 1.04 ha; brownfields, 5.05 ha; trees and scrub, 10.0 ha; wet woodland, 0.90 ha; rivers, 0.27 ha) as well as habitat not present but considered important (species-rich grassland, 23.69 ha; urban park, 1.67 ha; reedbed, 1.80 ha; ponds, 0.18 ha). Site-specific plans were developed for each habitat, based on existing UK Habitat Action Plans, and separate plans were created for targeted species of conservation concern within the area (Olympic Delivery Authority, 2008).

The ultimate aim of the BAP was to provide at least 45 ha of good-quality habitat by the end of 2014, and involved habitat installation during the construction phase, and the post-games (legacy) phase. Habitat creation involved ecological engineering techniques, for example the creation of living roofs and nest boxes on buildings, as well as the creation of habitat heterogeneity within individual patches. Where possible, habitat connectivity was considered, with corridors designed to connect habitat within the site, as well as consideration of the Park's location within the wider Lee Valley Regional Park. This redevelopment therefore represents a good example of urban conservation, with the incorporation of some key elements of urban and landscape ecology into its design, even though the effects remain uncertain. Annual monitoring of the site will be ongoing to determine the success of the initiative.

ecosystems within the park that were targeted towards specific species groups. Key stages in the process of ecosystem restoration are highlighted in Figure 7.1.

There are many studies of restoration and rehabilitation of urban ecosystems. A large proportion have focused on urban woodlands, and the ways in which they may be improved using techniques to reintroduce heterogeneity to the ecosystem and prevent seral development and dominance by a denser, less species-rich climax community. This is often achieved by re-creating disturbance regimes such as burning or thinning, or the removal of invasive plant species that may out-shade or out-compete native species. Such efforts may increase the range of species that can utilise these ecosystems, with, for example, structural heterogeneity allowing organisms of varying body size and habitat requirements to persist (e.g. Smith and Gehrt, 2010). In some cases reforestation is performed, but there is generally limited space available for this within the urban complex, and many trees planted are infrastructural or on private

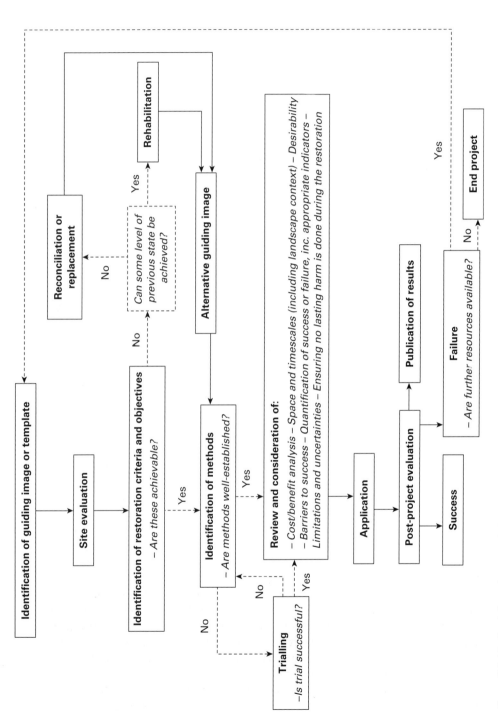

Figure 7.1 Idealised flow diagram of the process of ecosystem improvement via restoration, rehabilitation, reconciliation or replacement. Dashed arrows represent decisions that may guide the project development. Publication of results does not always occur but is highly desirable to inform further improvement efforts.

land, so do not contribute to specific reserves (though they do to the urban 'forest' in general, e.g. Bühler *et al.*, 2009).

The remediation and restoration of brownfield sites often takes place, with techniques such as soil-washing regularly applied (Box 7.1). This is often a precursor to redevelopment into the built environment, but sometimes into more palatable green space. There is evidence, however, that brownfield sites may be best left alone and that allowing natural regeneration to take place may be the best solution ecologically (Hartley *et al.*, 2012).

River restoration is also common in urban areas. As noted in Section 4.8, urban rivers present key ecosystems for improvement as well as the mitigation of disservices such as pollution and poor aesthetics. Rehabilitation can range from full re-creation of channel morphology and geomorphic elements, with planting of aquatic and riparian plants, to relatively small modifications such as channel widening, alteration of regulated flow regimes or the placing of in-stream structures to increase habitat complexity (Lake *et al.*, 2007). These techniques have had mixed success (see Lake *et al.*, 2007), and are often harder to quantify and evaluate because rivers are complex connected systems and difficult to manage in their entirety. Much restoration has focused only on individual reaches, rather than the entire system.

There are several salient themes that have emerged from the literature on urban restoration initiatives:

1   *Restoration is not trivial*, and because ecosystems are complex and species respond differentially to interventions, evaluating success in a meaningful way can be problematic (Hartley *et al.*, 2008). It is therefore important to establish appropriate objectives at the earliest stages, and determine how success will be measured. Much work has gone into the development of appropriate biological, physical or chemical indicators to measure restoration success. Biological indictors are the most ecologically meaningful, and usually focus on the presence/absence of sensitive species or particular assemblages, or in some cases more general increases in species populations or diversity (e.g. Purcell *et al.*, 2009). Often quantification is compromised because post-restoration recovery and population persistence requires time, and monitoring is not completed for long enough following the restoration action (Kondolf *et al.*, 2011).

2   *Isolated improvements can prove beneficial, but can have limitations.* There is increasing recognition that the landscape scale must be considered in ecological restoration, so that each restored area provides a complementary link within a broader network of habitat availability. As such, the creation of a more 'connected' network (see Chapter 2 and 8 for further discussion of connectivity) is more likely to maintain species metapopulations (e.g. Shanahan *et al.*, 2011), and consequently spatial planning becomes particularly important alongside the techniques applied at individual sites. Such connectivity may be maintained in a variety of ways: by creating corridors of green space; by managing the urban matrix (for example using reconciliation techniques as described below) so that there is increased similarity between the matrix and green spaces; and by ensuring the persistence of 'mobile links'. These are discussed further in Chapter 8.

3   *There is limited space available for restoration in urban areas.* This relates to both the general size of individual ecosystems (e.g. patches of land that may be used for

restoration) and the potential to create green networks in established urban areas. Although the establishment of such ecological networks remains a planning priority, all restoration must be conducted within the existing urban template of land use, and pressure remains high (and increasing) in most urban regions. Returning areas to a more natural state comes with an economic implication, as space cannot be used for other purposes. As a result, minor improvements and modification are most common, for example the improvement of existing patches of woodland or small river reaches. This limited space may be further complicated by different land ownerships, which may mean that different permissions and coordination are required for restoration initiatives. This is particularly problematic for river systems, which may have many riparian landowners or governing organisations, and makes coherent landscape scale management difficult. Consequently, it can be difficult to create patches large enough to support populations, and to establish sufficient connectivity to support metapopulations.

4   *The end result may not be what is expected.* Ecological restoration remains an emerging science, and the varying complexities involved means that the characteristics of the 'restored' ecosystem may not be entirely as planned. For example, assemblages and seral trajectories may not proceed as anticipated and in part this is dependent on remnant environmental conditions, soil seed banks, the success of reconstruction, pre- and post-restoration assessment of the ecosystems, and trends of colonisation by species in the area (e.g. Pavao-Zuckerman, 2008).

5   *Species reintroductions are relatively rare in urban conservation.* The reintroduction of species that have been lost within a region represents another form of restoration as it is an attempt to re-create a past condition or assemblage. Urban reintroduction usually focuses on plant or invertebrate species (e.g. Hannon and Hafernik, 2007) as they are species that often require relatively limited habitat and are easy to manage. Usually reintroduction is associated with other forms of ecological improvement. Success is often limited as conditions may not be sufficiently improved to allow successful re-establishment of populations or maintenance of population viability over the longer term. Hannon and Hafernik (2007), for example, recorded a successful reintroduction of the damselfly *Ischnura gemina* into a restored river channel in an urban park in California, USA. However, establishment did not persist beyond the second year, as the initial clearance of overgrowing vegetation in the river channel that facilitated reintroduction was not maintained, and so suitable habitat was subsequently lost. Such initiatives usually focus on a limited number of species, and unless they are keystone species may have little influence on biodiversity or ecosystem processes. Passive restoration, which is where environmental conditions are improved and recolonisation occurs naturally, is generally a better alternative.

6   *Societal factors are particularly influential in urban restoration.* Often ecological improvements are performed not just to increase diversity and functioning but to improve public use or appreciation of a site. The best restoration schemes usually involve a range of stakeholders, from scientists to the local community, and have objectives that cut across ecology and society (e.g. Kim, 2005). This can, however, be difficult to manage and can create tensions, with conflicting priorities and objectives. Important societal considerations include what people want from a restoration, what they will value and therefore look after, the level of

understanding needed, and their perception of, e.g. aesthetics (e.g. Junker and Buchecker, 2008). In many cases of urban restoration, societal benefits outweigh the ecological. Miller (2006) records how land acquisition and restoration of prairie grassland in the urban region of Kane County, Illinois, would probably be insufficient to support populations of grassland birds due to the limited area available; but that the societal benefits of small prairie areas are substantial. They may help to allow urban residents to re-engage with species and ecosystems that they would not otherwise readily experience, develop an appreciation for species and conservation, and mitigate the 'extinction of experience' and 'generational amnesia' common to those living in urban areas (Miller, 2006).

7.4 Reconciliation ecology

As noted above, the idea of reconciliation ecology is essentially that anthropogenic ecosystems can be ecologically improved without losing their capacity to be used by society – indeed, may even enhance it. Some of the techniques used in reconciliation may be similar to those in restoration or rehabilitation. The three main differences are:

1   Whereas restoration/rehabilitation efforts use a previous state or unimpacted template to determine objectives and evaluate success, reconciliation uses other measures, such as a more general increase in species richness/diversity, presence of sensitive or desirable species or increase in ecosystem function. For example, the inclusion of a wider range of habitat types in an urban park may be considered reconciliation – the park is still being used by citizens as intended, but more species can utilise it.
2   The built environment can be enhanced without any formal 'restoration' and therefore the applications are much broader in potential scale, even if more limited in what they may achieve at local scales.
3   Reconciliation efforts are driven much more at the local scale, from individual or community group actions rather than by local government investment. This does not mean that governmental or top-down approaches are not conducted, however; the planting of infrastructural trees for example is often managed by local government, and in fact a spectrum of reconciliation activities exist, driven at different levels (Figure 7.2). In some cases actors operate across the different scales – for example, government agencies may encourage local action by creating planning incentives or requirements for living roofs (Carter and Fowler, 2008).

The reconciliation concept is perhaps the most relevant to urban conservation, and the most realistic option in many urban environments. It clearly relates to the realities of urban environments, including the dominance of recombinant assemblages and novel ecosystems (see Section 6.4 and Section 7.7). Reconciliation ecology is consequently not just a broad concept, but an important ecological issue of relevance for enhancing diversity, improving ecosystem services and maintaining species populations indefinitely. This form of ecological engineering is in its infancy (some elements include living roofs and walls, discussed in Chapter 5), but there are again some emerging themes:

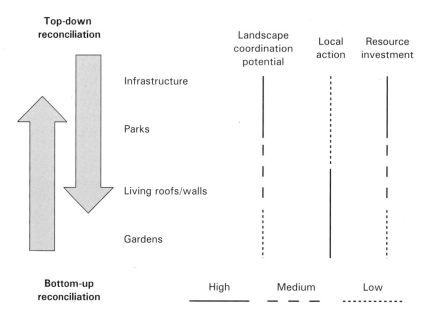

Figure 7.2 Possibilities for urban reconciliation ecology, both led by local/regional governments (top-down) and led by conservation citizenry (bottom-up). Improvement of built infrastructure and public green space requires support by local or regional government or other organisations, and may be better coordinated at broader scales. These initiatives are also more likely to require resource investment. Improvement of private green or built space (roofs, gardens) may be performed by motivated citizens, but is harder to plan and coordinate. They may require lower levels of resourcing. (Reproduced from Francis and Lorimer, 2011, with permission.)

1   *Urban regions are particularly relevant to reconciliation ecology.* There are various ecological engineering techniques being developed that allow 'greening' of the built environment, as well as improvement of existing green space. These range from the construction of living roofs and walls to various wildlife gardening techniques to full eco-city design (e.g. Lundholm and Richardson, 2010; Francis, 2011). Given the abundance of the built environment, 'reconciling' it with other species seems the most logical way of providing increased habitat area to a range of species.

2   *Reconciliation is most likely to have an impact at the landscape scale.* Although there has been limited research into reconciliation techniques, it has been noted that local-scale interventions, while useful, are most likely to contribute to the maintenance of metapopulations, and ultimately biodiversity, if coordinated at the landscape scale. This is essentially the principle of making the urban matrix (see Box 2.1) more like the patches of green space that may be found in the urban landscape, thereby increasing habitat area and decreasing isolation.

3   *The end result may not be what is expected.* As with restoration, it is not always possible to create habitat for a specific species, and reconciliation may not be the best option for encouraging populations of rare or endangered species. This form of habitat enhancement may well lead to 'more of the same' species generally found in urban environments. Ideally, reconciliation would also either reduce

habitat for pest species, or at least not provide more, though this may be difficult to achieve in practice or over large scales (Lundholm and Richardson, 2010).

4   *Reconciliation is not a substitute for preservation.* Although reconciliation is likely to have a supportive effect on urban biodiversity, it will not create the same extensive, contiguous habitat with remnant environmental conditions (e.g. soils, seed banks) that ecological reserves may contain. Consequently, reconciliation should ideally be used as a way of supporting existing green space where possible.

5   *Reconciliation may be retrospective and pre-emptive.* Reconciliation activities can be retrospective, wherein engineering is applied to existing built environments (such as retrofitted living roofs installed on existing buildings); or pre-emptive, wherein building designs can incorporate ecological engineering (e.g. living roofs or other greenery) based on ecological principles from the start. Pre-emptive efforts are more desirable than retrospective, but this remains an emerging area (see Chapter 8).

Some examples of reconciliation are now considered in relation to green space and the built environment.

7.4.1 Reconciliation ecology in urban green space

Urban green spaces offer abundant opportunities for reconciliation ecology. As they are usually utilised for recreation or other societal benefits, modifications to increase biodiversity are relatively easy and unlikely to compromise their societal use. Public parks and private gardens are perhaps the best candidates for reconciliation activities.

As urban parks are already vegetated to a greater or lesser extent, there is notable potential to enhance habitat quality to support larger numbers of species. Lovell and Johnston (2009) suggest that parkland and any connecting corridors can be planted with species intended to provide food sources for herbivorous fauna with low urban populations, or provide habitats that are uncommon in urban areas, for example the replication of heathland or savanna (Gobster, 2001). Habitat creation, creative conservation and landscaping for biodiversity are now becoming common in urban parks (e.g. Gobster, 2001). Even in heavily used parks and recreational areas, more modest kinds of sympathetic management for biodiversity may include the planting of wild flower gardens or grassland areas, trees (including snags) that support the foraging or nesting of specific bird taxa, or establishing vegetated and tree-lined buffer areas along streams or other natural corridors within parks to support small mammal populations (Mahan and O'Connell, 2005; Sandström *et al.*, 2006).

Such habitat modifications are easily compatible with both recreational and educational uses and can be performed over larger spatial areas and potentially with greater coordination in urban parks and other public recreational areas than on living roofs, walls, and private gardens. However, as most parks and recreational spaces are public owned, this larger scale of reconciliation will require a different level of commitment and has less potential to be driven by the local actions of interested citizens. Nevertheless, parks may represent a nucleus for reconciliation techniques and biodiversity support that local action can enhance, for example by garden management (Colding, 2007; Forman, 2008).

Gardens are highly variable in their size, shape, management, habitat provision and position within the landscape and offer good opportunities for reconciliation

(Goddard *et al.*, 2010). Gaston *et al.* (2005b) conducted experimental tests of habitat improvement methods in urban gardens and found that the addition of simple artificial nest sites, small plastic ponds and patches of nettles encouraged the presence of a range of invertebrates, though some additions were more effective than others. For example, artificial nest sites such as tin cans, wooden blocks and plastic pipes were very successful in encouraging nesting of solitary bees and wasps, while pots and boxes introduced for bumblebee nest sites did not encourage any nesting of these taxa (Gaston *et al.*, 2005b). For wide-ranging species, the addition of unusual habitat types, less common food plants, or nesting sites may perhaps be the best method for encouraging viable populations of the species, for example shrew and mouse boxes, owl boxes, rock piles that may be used for dens, etc. Ideally, such habitat enhancement techniques should be conducted within the context of the wider environment that the species will use (see Goddard *et al.*, 2010).

Gardens are generally owned or managed by private individuals for recreation, and this ownership structure, while meaning that reconciliation activities are easy to implement, also presents some challenges. For example, some attempts to encourage 'wildlife gardening' techniques have met with resistance from garden owners due to a reluctance to allow a garden to seem too wild or unmanaged (often linked to the possibility of disapproval from neighbours; Gaston *et al.*, 2005a). Although many people like to view wildlife in gardens, they may be unwilling to utilise techniques that they see as being in any way unappealing or unattractive. An example of this is the potential for nettle (*Urtica dioica, U. urens*) patches to be used to encourage a greater range of invertebrate species to utilise garden patches, but as nettles are unattractive and have an irritating sting, garden owners are generally reluctant to have patches of sufficient size to make a difference to species populations in their gardens (Gaston *et al.*, 2005b). Similar problems have been identified in efforts to persuade people to keep dead wood in their gardens.

There is also the challenge of coordinating reconciliation efforts in a heterogeneous and highly fragmented garden landscape with multiple and different land managers. So far there has been little work conducted into how gardens may maintain species populations or metapopulations at the landscape scale and over longer (e.g. decadal) temporal scales, and the implications this may have for diversity, though garden areas and habitat are increasingly being incorporated into urban landscape planning (Jim and Chen, 2003). The recent popularity and success of large-scale media wildlife programmes focused on domestic gardens – like the RSPB's Big Garden Bird Watch and the BBC's Springwatch in the UK – suggests that there might be great potential for coordinating citizen gardening activities. However, domestic gardening activities can have a negative effect on biodiversity, for example acting as source patches for the spread of non-native species into the wider environment, or hunting grounds for domestic animals that are detrimental to many species, e.g. the domestic cat (*Felis catus*) (Goddard *et al.*, 2010).

Reconciliation in gardens therefore requires some level of enhancement across a suite of gardens rather than on an individual level. As with the species–area relationship found for gardens (Section 4.3), the enhancement of garden habitat that is beneficial for specific taxa is also likely to have a cumulative benefit for populations and diversity at broader spatial scales, even if the capacity for increasing individual garden sizes is low in most cases.

Wherever possible, reconciliation efforts should aim to reduce any negative impacts associated with gardens as well as enhancing habitat, though there may be much individual opposition to controls on, for example, non-native planting or cat control. Given their abundance within urban regions and their private ownership, the promotion of general habitat heterogeneity, structural complexity and the creation of specific habitats to support targeted species in domestic gardens is a real possibility for widespread urban reconciliation efforts, and is likely to feature heavily in the future of urban conservation.

7.4.2 Reconciliation ecology in the built environment

Perhaps the greatest potential application of reconciliation ecology is on the built environment. The utilisation of living roofs and walls for reconciliation has been discussed by Francis and Lorimer (2011), who suggest that there are a wide range of habitats that can be created on roofs and walls, which may support a substantial range of species. This may particularly be the case where the opportunity exists to allow 'extensive' or relatively unmanaged roofs to be installed, and where the provision of a range of habitats or microsites (e.g. different sediment types and depths, piles of wood or stones) is possible (see, e.g. Bates *et al.*, 2009). Much work has been conducted at very local scales, such as individual roofs or comparisons between a small number of roofs, and it remains to be determined whether such installations may support species metapopulations at the landscape scale. As for gardens, societal factors such as cost and aesthetic preference may pose barriers to widespread use of such techniques, though greened buildings are generally considered pleasing and desirable (e.g. White and Gatersleben, 2011).

Transport infrastructure also offers some opportunities for reconciliation. Roadside verges may be enhanced for biodiversity during construction and post-construction maintenance by preserving any original vegetation and soil, preserving the seed bank, keeping nutrient availability low to prevent competitive dominance, maintaining variable topography and soil moisture to create fine-scale heterogeneity of environmental characteristics, planting species with different morphologies and functional characteristics, incorporating mowing or other disturbance management to keep competitive species from become too dominant, and allowing additional rooting zones for trees under new roads (e.g. Harper-Lore and Wilson, 2000; Valtonen *et al.*, 2007). There is abundant literature on the use of underpasses to prevent roads from acting as barriers to species movement, and although this does not create habitat for species, it may allow increased utilisation of other urban habitats by species that may experience roads as substantial barriers (Beben, 2012), and may thereby support broad-scale reconciliation efforts.

Possibilities for habitat enhancement may exist via the spatial arrangement of urban trees, for example the creation of continuous tree lines or increasing tree densities, though the benefits of such approaches remain to be determined. Species may also be selected to provide certain habitat features, for example as roosting sites for bird or bat species, or conversely, as a method for reducing populations of urban generalists that have become pests (see Box 6.1). Establishing the benefits of planting specific species for urban biodiversity, alongside aesthetic or management considerations, is a further research area for urban reconciliation ecology.

7.5 Conservation governance

Conservation decisions and actions are governed at a range of socio-political levels, and such governance is central in determining the types of conservation initiatives that are performed. While legislation may be set at international levels (Table 7.2), further decisions and actions cascade down from government to agencies operating at national or regional scales, and thence to local organisations that may enact the policy using professional staff and/or a volunteer workforce. There has been much discussion of the implications of this for conservation, highlighting gaps in knowledge or training (Newing, 2010), partisan preferences for particular species that may be driven by political or social agendas or simple lack of understanding (Lorimer, 2006), and geographical disparities in process between (for example) local governments (Evans, 2004), among other things.

Consequently, not only is the science of conservation difficult, determining what is needed and at what scale, but also the issue of governance is complex. This has led to the battlefield concept of 'triage' being applied to conservation action (Bottrill *et al.*, 2008). The principles of triage in this context hold that it is important to establish priorities based on level of threat, ecological value of species/ecosystems under threat, cost/benefit of conservation interventions, and what is feasible to practically achieve (Figure 7.3). Following initial implementation, progress should be evaluated and

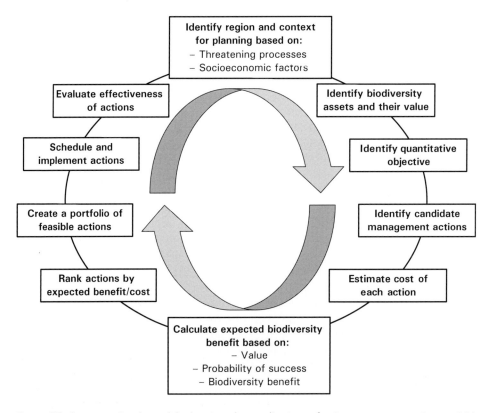

Figure 7.3 An operational model showing the application of triage to conservation activities. (Reproduced from Bottrill *et al.*, 2008, with permission.)

further priorities modified accordingly, for example based on success or lack of it, unexpected complications or emerging threats, and so on. Within this wider context, urban ecosystems may be considered a low conservation priority compared to some threatened and highly biodiverse ecosystems, such as those found in the biodiversity hotspots of the world (e.g. Joppa *et al.*, 2011). As noted earlier, however, their contribution to the functioning and metabolism of the wider urban region and their ecological and societal services and benefits may be substantial. Vast numbers of people may be directly exposed to them; as a result, people may 'buy in' to urban ecosystems and ecology much more in the future, especially as their importance becomes increasingly recognised.

Francis and Goodman (2010) have argued that conservation is a form of post-normal science, wherein a range of stakeholders may have useful roles in contributing to conservation decisions and actions (Figure 7.4). In this 'extended peer community' of stakeholders, governmental organisations, ENGOs, professional consultants,

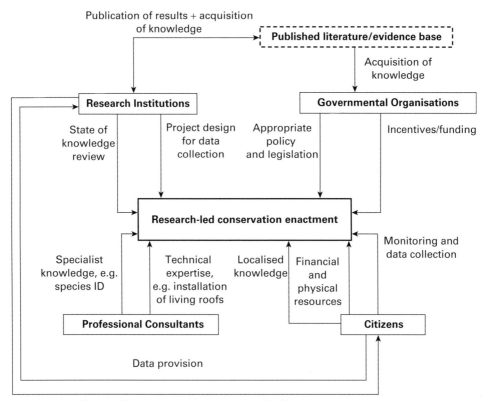

Figure 7.4 A model showing idealised linkages between different components of an extended peer community for more effective conservation practice. Citizens are the key group within the model, driving much of the conservation activity, with other members of the peer community providing key knowledge, advice, services and incentives. Ideally an evidence base (e.g. via publication or creation of an online database) is created as an output alongside the conservation itself, to inform activities elsewhere. (Modified from Francis and Lorimer, 2011; reproduced with permission.)

universities or research institutes, and the citizenry would ideally interact to review, revise and validate conservation practice. In reality, much conservation practice does not have this level of engagement, which may lead to limited progress. This is recognised in the conservation community, and there are increasing efforts to bridge gaps between community members. For example, Kondolf *et al.* (2011) report how the use of students to conduct post-project appraisals of river restoration projects, which are generally lacking, has created a database of over 300 studies that are available online. Practitioner-based online journals such as *Conservation Evidence* have been developed to encourage dissemination of best practice and emerging issues amongst the extended peer community. Urban areas may represent important crucibles for the development of more synthetic forms of conservation governance, for some of the reasons noted in Section 1.7. Although their conservation may be viewed as a low priority, the ecosystem service benefits of urban ecosystems in particular may help to stimulate investment. The human workforce, expertise (amateur, professional and academic), technology and finances that are concentrated in urban regions, alongside the relaxation of 'expectations' (e.g. preservation of rare or endemic species) of urban ecosystems that may be found in more proscribed conservation programmes for natural systems, may mean that urban ecosystems emerge as exemplars of effective conservation in due course. Increasing awareness of urban ecosystems is key to this.

Research for validation of conservation practice is a major issue, and this was one reason for the instigation of more rigorous 'systematic reviews' following the medical model to determine effectiveness of applied techniques (see Pullin and Knight, 2009). Adams (2005) notes that research was a limited aspect of many programmes driven by state wildlife agencies in the USA until the start of the twenty-first century, and as such this has helped justify the creation of the Long-Term Ecological Research (LTER) sites (see Box 1.1 for the urban LTER sites at Baltimore and Phoenix). In reality, most conservation is performed by amateur or semi-professional groups, and is guided or advised by more professional groups. Much urban conservation has been driven by local government or ENGOs acting locally, a trend that is set to continue (e.g. Adams, 2005), particularly for fine-scale improvements resulting from reconciliation ecology. Urban areas do have one particular strength for building an evidence base for conservation, and that is the abundance of interested citizens that are on-hand to work in urban ecosystems. A good model for mobilising and utilising this resource is 'citizen science'.

7.6 Citizen science for urban conservation

All forms of conservation rely to some extent on the public, whether for funding, approval or implementation. The question of data collection and validation to demonstrate that a conservation activity has a beneficial effect is a difficult one. It makes sense to apply the motivated citizen workforce to this gap, and this is increasingly being incorporated into research models (Cooper *et al.*, 2007). Citizen science is essentially the conducting of scientific research or collection of data by amateurs, in collaboration with professional researchers (e.g. Cooper *et al.*, 2007). Usually citizen science is applied over broad geographic areas and long timescales, at scales where dedicated professional research would be too expensive. Amateurs are usually trained by researchers and samples of data collected are validated by professionals, to ensure standardisation.

The model has been employed in a range of urban contexts, including monitoring of trends of cat predation on urban birds (Cooper *et al.*, 2012), butterfly diversity (Matteson *et al.*, 2012), microsite use by urban bees (Everaars *et al.*, 2011), human interactions with coyotes (Weckel *et al.*, 2010) and garden use by mammals and birds (Cannon *et al.*, 2005; Toms and Newson, 2006). The range of non-urban uses is even greater, and volunteers are becoming increasingly important for collecting ecological data (Silvertown, 2009). Some urban citizen science initiatives can be extensive: The Open-Air Laboratory Project (www.opalexplorenature.org) in the UK, for example, has engaged with citizen science to increase public engagement with the environment at a range of levels, and with a cost of £12 million over 5 years (Silvertown, 2009). The project is partly based around extensive surveys and data collection relating to soils, air, water and biodiversity; many of the surveys have understandably had great success in urban environments.

The citizen science model is particularly appropriate for urban areas (e.g. Cooper *et al.*, 2007) and has been suggested as a mechanism for the widespread urban application of habitat creation and improvement efforts along the lines of urban reconciliation ecology (e.g. Goddard *et al.*, 2010; Francis and Lorimer, 2011). Several studies have established that given appropriate training and data validation, citizen scientists can be highly effective and reliable, though sampling and reporting/validation protocols need to be rigorous to ensure consistent sampling quality and effort (Cooper *et al.*, 2007; Matteson *et al.*, 2012). Cooper *et al.*'s (2007) 'Adaptive Citizen Science Research Model' is perhaps the best model for evidence-based urban conservation. In this model, both ecological (e.g. habitat provision) and social (awareness of invasive species, greater appreciation of urban ecosystems and ecosystem services, environmental stewardship) improvements are attempted by the citizenry and their success measured. Information on success (or lack of it) then feeds back into further initiatives and management strategies, so that strengths can be identified and built upon. This model therefore gives greater responsibility and inclusivity to citizens as it moves beyond data collection to ecological and social manipulations. The wider sharing and reporting of data and successful initiatives, as well as the input of other elements of the 'extended peer community', including academic, governmental and non-governmental organisations helps to capitalise on successful improvements as well as dispel 'extended facts' (see Francis and Goodman, 2010) that may have limited truth or that are not statistically conclusive. Initiatives such as this are required for the successful implementation of urban conservation, and are a further reason for greater encouragement of citizen engagement and interest in urban ecosystems and ecology.

7.7 Moving towards novel ecosystems and assemblages in conservation viewpoints

Novel ecosystems and their 'recombinant' species assemblages are characteristic of urban environments, and there is a growing acceptance that such systems will become increasingly abundant across most of the globe in the coming centuries (Hobbs *et al.*, 2006; Meurk, 2010). Certainly there is a body of ecologists who consider that such a trend is unavoidable, and that the usual conservation paradigms have to be reconsidered in relation to these changes (Hobbs *et al.*, 2009). As Meurk (2010) states: 'we have to go with the flow rather than fighting nature head on, make the best of it and derive something useful' (p. 215). This strikes to a consideration at the heart of conservation: biodiversity loss is essentially a human concern, and is a threat to the existence of human civilisation, which has emerged over the last 10,000 years and in the context of a relatively stable (until recently) environment. Our attachment to 'natural' ecosystems is a product of both the environment that encloses our cultural history and the recent reconceptualisation of 'nature' and 'wilderness' from a 'removed' urban perspective. This may perhaps be a constraint that should be reconsidered with an objective eye, and 'novel ecosystems' represent a step towards this. It may be argued that there is relatively little merit in preserving rare species or re-creating 'natural' or pseudo-natural ecosystems in urban environments without strong justification, but improving systems and their functioning is always a worthy and practical ideal. However we may move forward with nature conservation and addressing the biodiversity crisis, novel ecosystems are the future and should be accepted as both an inevitability and an opportunity. Urban ecologists are amongst those at the forefront of this perspective, because of the novel nature of many urban ecosystems. As Francis *et al.* (2012) note:

> urban ecosystems should be viewed as young, emergent systems from which we can learn much about ecosystem processes and the role of humans in ecology; the opportunity to study such nascent ecosystems and biotic assemblages should not be lightly declined.
>
> (p. 186)

Lundholm and Richardson (2010) consider that distinctions can be made between novel and 'analogous' habitats, depending on particular ecosystem characteristics, level of human influence, and the range and types of species that may use the habitat. For example, artificial or heavily contaminated urban soils, sewerage systems, recombinant assemblages and new hybrid species may all represent novel urban habitats for different organisms, created by humans, which are not found elsewhere in nature. Built surfaces, lawns, constructed wetlands and salt-rich roadside verges and ditches may create analogue urban habitat that echoes that found in nature, such as cliffs or talus, semi-natural grassland or salt marsh. Much of this distinction relates to what is considered 'novel'; for example an artificial surface is still somewhat novel even if it is analogous to a natural habitat, and there is probably a continuum of novelty in the urban environment (Lundholm and Richardson, 2010). Further quantitative work is needed to compare urban ecosystems and the habitat they may represent both within and between urban areas, and between urban and natural areas. This quantification will form a good basis for future conservation in urban regions. In general, urban conservation will probably continue to contain elements of preservation, restoration and reconciliation, but will move towards an acceptance of more varied, novel systems and assemblages, and the improvement of both system functioning and sustainability over preservation of particular species/assemblages.

7.8 Chapter summary

- Contrary to uninformed opinion, nature conservation is highly relevant to urban ecosystems, particularly because of their close association with humans and the benefits that may result from urban conservation efforts.

- Conservation may be based around the reservation, restoration and reconciliation of species and ecosystems. All of these have relevance for urban ecosystems, though the ecological engineering techniques associated with reconciliation ecology may have the most wide-ranging applications to both green space and the built environment.
- All forms of urban conservation need rigorous design and monitoring for evidence of success or failure, to form a basis for further efforts.
- Conservation governance is complex, particularly in urban areas where many agencies may be involved. Conservation may be considered a form of post-normal science that involves a range of stakeholders (governmental organisations, ENGOs, professional consultants, universities or research institutes, the citizenry) that should be involved in decision-making and implementation.
- Urban citizens represent an under-utilised resource that can be brought to bear on urban conservation initiatives, particularly those within the sphere of reconciliation ecology, and an adaptive citizen science model may be the best method for achieving this.
- Urban conservation will increasingly be shaped by an awareness of urban ecosystems as 'novel ecosystems', and by the desire to create functional and sustainable assemblages and ecosystems within urban environments, rather than preserving or re-creating traditional 'pre-urban' conditions.

7.9 Discussion questions

1   How useful is the distinction between the different forms of conservation? Can you think of some examples of each from your urban region? How do they differ?
2   Consider an urban river that has experienced an improvement in water quality. Vegetation is subsequently planted, and fish species found within the pre-urban system are re-introduced. Would you consider this to be restoration? Why?
3   Make a list of organisations you know that are involved in urban conservation. Are there many? Are they governmental or non-governmental? What is their remit? How successful are they?
4   List the benefits of citizen science for urban conservation. What obstacles might exist to the successful implementation of such a model for urban conservation? If you were to design a protocol for an urban conservation problem, what would you need to consider?
5   How valid is the 'novel ecosystems' concept? Do you agree that supporting recombinant assemblages is better than preserving particular communities associated with more natural ecosystems?

Chapter 8

Incorporating ecology in urban planning and design

8.1 Introduction

Consideration of urban ecosystems and their ecology in urban planning and design is important for ensuring that urban regions maintain as high an environmental quality as possible. This in turn means that they may be more sustainable, with an enhanced capacity to provide ecosystem services, as well as making them more pleasurable environments in which to reside. There has always been an element of integrating nature into urbanisation, but as both urbanisation and ecological awareness have increased in recent decades, the objectives, principles and techniques applied to urban planning and design have changed. Urban planning usually relates to longer-term strategies and policies relating to land-use change across broad spatial areas, while design is short-term, fine-scale implementation of land-use change or modification, though the two are clearly interrelated and in many cases the terms are used somewhat interchangeably. Both are considered in this chapter.

The intentional preservation of vegetated areas from urban development, and in some the cases the specific creation of greenery in semi-arid areas through manipulation of water resources, has taken place since antiquity. Persian and Roman cities, for example, had both private and public parks and gardens (Gleason, 1994; Faghih and Sadeghy, 2012). These early green spaces were not created for ecological purposes but rather for recreation, aesthetics, to provide water and shade, for religious or spiritual matters, or sometimes in order to demonstrate power and wealth. These broad aims have generally remained the focus of urban planning and design, and it is only relatively recently that ecological considerations have emerged. For example, the urban parks movement that began in the nineteenth century and led to the development of major urban parks around the world (Harnik, 2006) was instigated out of a desire to improve the living environment for citizens, rather than to safeguard ecosystems or species – such concerns not having entered the public consciousness at that time. Likewise, the notion of 'garden cities' espoused by Ebenezer Howard at the end of the nineteenth century (Howard, 1898), involving the creation of defined urban regions with planned green space and contained within a green belt, was aimed at melding rural and urban ideals to create pleasant and healthy living environments for citizens.

Even some of the most significant planning instruments that have helped to preserve green space within urban regions were not specifically intended to have ecological benefits. An example is the green belt, which has been a major tool for urban planning since the late nineteenth century. This is a contiguous boundary of green space

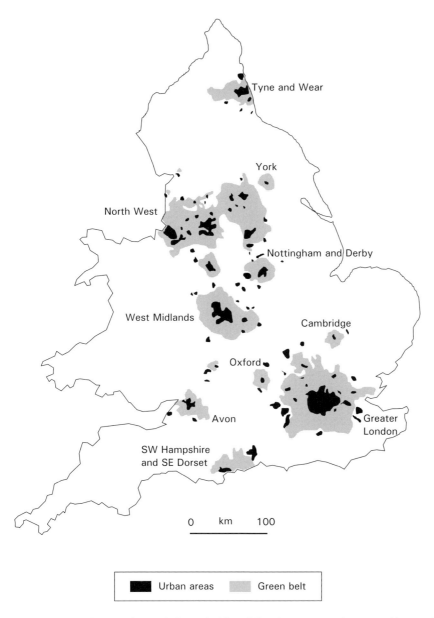

Figure 8.1 Distribution of green belts in the United Kingdom, surrounding several large urban regions. (Redrawn from Amati and Yokohari, 2006, with permission.)

encircling an urban complex, usually composed of natural or semi-natural vegetation mixed with agricultural land and commercial forest plantations. The green belt concept was proposed principally to contain urban expansion, and to ensure a distinction between 'rural' and 'urban' regions (Amati, 2008). As a formal planning tool, it was first developed in the UK (Figure 8.1), though the principles of preserving exurban green space were established in many other locations (e.g. across Europe and the USA) in the early twentieth century (Ignatieva *et al.*, 2011). The UK model was applied throughout the world, with varying success, until the 1970s. By this time the limitations of green belts in halting expansion had become more apparent. As with most green space, green belts were also intended to provide opportunities for recreation, particularly where green space was lacking within the urban complex itself. They were certainly not designed to preserve space for biodiversity (this mainly being a recognised concern from the 1980s onwards), though they have since become important for many species as surrounding areas have been degraded. This is reflected in the observation that much of the land used for early green belts consisted of former estates, farms, common land and golf courses (Amati, 2008) rather than land that was ecologically 'valuable' or natural, and also in the gradual urban encroachment that often occurs on green belts (Amati and Yokohari, 2006). More recently, planning for green space and urban ecology has questioned the divide between urban and non-urban spaces, and has followed the path of spatially integrated 'greenways', 'green corridors' or 'green networks' that weave through the urban region rather than encase it. Much of this has developed from the principles of landscape ecology considered in Chapter 2 (particularly landscape connectivity), and is considered here in more detail.

Recognition of the ecological benefits of urban green space is a recent development, as is the incorporation of green space preservation or restoration in urban planning specifically to support biodiversity and urban sustainability. Despite an increasing focus on ecologically sensitive urban development, the design of green spaces and networks, and the potential for ecological and environmental engineering of urban environments, rigorous spatial planning for urban ecology remains a nascent field for both implementation and scientific investigation. This chapter first considers the role of planning in ecologically sensitive urbanisation that aims to minimise biodiversity impacts and preserve green space. It then considers the incorporation of ecological principles into green space planning and design, and the spatial planning of green networks. Finally, there has more recently been an awakening of the possibilities of ecological and environmental engineering, particularly of the built environment, in order to provide habitat for species within the urban environment, but also to mitigate some of the environmental problems that plague urban ecosystems. This aspect of urban planning and design is also briefly considered.

8.2 Planning for ecologically sensitive urbanisation

Increasing recognition of the importance of urban ecosystems for biodiversity and ecosystem services, has led to the development of varying instruments to allow for ecologically sensitive urbanisation. These may be legislative (i.e. legally binding) or advisory (providing guidance on a course to action to achieve desired objectives). Specifically, many countries have legislation relating to biodiversity conservation, often linked to global or regional conventions or directives such as the Convention on Biological Diversity (see Table 7.2), and specific planning legislation relating to both biodiversity and the wider environment, which may incorporate (for example) flood risk management, air quality and green space provision (e.g. Sadler *et al.*, 2010). In total, these may translate to the preservation of urban green space or other forms of ecologically sensitive urbanisation. Legislation is usually more effective than policy-based instruments alone (e.g. Jongman *et al.*, 2004), though planning policy at the highest levels may also greatly work to reduce impacts from urbanisation. For example, decisions made on whether to adopt a 'compact-city' approach to planning, and therefore support densification rather than expansion, may help to prevent exurban ecosystem fragmentation (Jabareen, 2006; Catalan *et al.*, 2008). The designation and enforcement of planning zones, common in most developed countries, may also regulate development and allow it to be guided towards specific areas to protect target ecosystems or minimise certain types or areas of expansion (Masek *et al.*, 2000). Zoning is usually based on land-use type or activities, and may regulate building type, height, density, infrastructure, impervious surface cover, and so on; in some cases zoning specifies green space requirements and compensation measures required for development (e.g. Yong *et al.*, 2010).

Legislation that protects individual species of conservation importance, such as the Wildlife and Countryside Act (1981 and supplements) in the UK, and the Endangered Species Act of 1973 in the USA, may prevent urban development or require certain types of mitigation to allow development to take place. Mitigation measures vary, and there is a 'mitigation hierarchy' that is usually followed in any given development. Mitigation usually takes place after the determination that one or more species of conservation concern will be detrimentally affected by a planned development. The hierarchy consists of:

1   avoidance of impact, i.e. developing an alternative area or using an alternative design that will not create the impact;
2   minimising impact, for example by conducting development during periods when species are least likely to be disturbed, or retaining key habitat features within the developed area;
3   restoration of biodiversity or habitat, whereby habitat is re-created in some form on the developed site, and species populations reintroduced as required.

If these options are not possible on the site, often some compensatory mitigation, or the creation of habitat elsewhere (also termed 'biodiversity offsetting') is sometimes required (see Kiesecker *et al.*, 2010). The hierarchy may be difficult to implement in some cases due to absence of data on impacts or species distributions, or subjective judgements of the scale of impact or appropriate response (Kiesecker *et al.*, 2010). It is nevertheless a staple of urban development in many countries. One potential problem within the urban context is that it can be difficult to justify mitigation if the urban ecosystems being developed do not support any species of conservation concern, and are therefore not legally protected.

It is increasingly common for proactive mitigation measures to be applied to urban development. These are often linked to ecological engineering techniques (see Section 8.5). Proactive mitigation follows the principle that any development should provide ecological habitat as part of its design, and therefore chimes with the principles of reconciliation ecology (Section 7.4). Importantly, the incorporation of ecology within the design itself is more likely to achieve completion and success, rather than occurring as an afterthought or token gesture. Xu *et al.* (2011) demonstrate this for the construction of 'near-natural greenways' in China, which is essentially the ecologically sensitive development of road and rail infrastructure. This case included:

1   the construction of wildlife tunnels to allow animal movement and migration;
2   the removal of soil (with seed banks) and vegetation from sites prior to development, so that it could be reinstalled once building was complete;
3   the avoidance of wetlands and other sensitive ecosystems that may be particularly impacted by the development;
4   the use of only native species in roadside verges, to minimise the risk of the spread of non-native and invasive species.

These measures are implemented alongside more general environmental provisions of waste control, cleaner energy for tunnel lighting, and so on. This type of sensitive design may be increasingly important in expanding urban regions, where both extensive and densely arranged road networks are being constructed. This particular example included both habitat avoidance and provision in its design, but in many forms of urbanisation avoidance is the less feasible option. Consequently, habitat provision takes precedence; often using ecological engineering techniques such as living roofs (see Sections 5.3 and 8.5; Box 7.1).

A major obstacle to ecologically sensitive urban planning has been the provision of suitable types and quantities of data and their synthesis to allow accurate decision-making. In many cases a professional survey of a planned development site and its immediate environs is conducted to determine the presence of species of note, often as part of a wider Environmental Impact Assessment (EIA) required by law (Mörtberg et al., 2007). In recognition of the importance of understanding and managing urban biodiversity at the landscape scale, there are increasingly sophisticated tools being developed for decision-making related to urban planning. These often incorporate species and habitat distribution databases with socio-cconomic databases (e.g. relating to suitability for building, population density, amenity values, etc.) that are measured and visualised using spatial tools such as geographical information systems (GIS), which may be applied at varying spatial scales to create a more informed judgement on a proposed development. Despite some early spatially nested decision-support systems (Box 8.1), they remain rare and most decisions are made at relatively local scales. They are further compromised by generally being developed for individual species or sometimes groups, which may not accurately reflect wider patterns of biodiversity. This may be mitigated to some extent by focusing models on more sensitive species, with an assumption that planning and design that safeguards more vulnerable species should maintain many less-sensitive species. Figure 8.2 shows how such a model may be developed (Mörtberg et al., 2007).

As an example, Mörtberg et al. (2007) developed a landscape ecological assessment (LEA) methodology to predict impacts on habitat availability and biodiversity in urban development scenarios. They examined three scenarios for Greater Stockholm in which 250,000 new residences were being created over 30 years: densification, diffuse expansion and expansion directed along transportation corridors. To determine the ecological impacts of these scenarios, they were considered in relation to two important biological conservation targets: (1) preservation of large forested areas with associated wetlands, mainly found in exurban or complex-edge areas; and (2) preservation of forest fragments located near to residential areas, mainly within the urban complex. Impacts were based on likely changes in habitat provision under the three scenarios for three focal forest-dependent species. These were capercaillie

Box 8.1 **Landscape planning for ecologically sensitive urbanisation in Colorado, USA**

Colorado is a key area for residential development, but there is also an awareness that habitat has to be preserved. In the late 1990s, population growth was around 3 per cent, and there was a concern about habitat loss in the Rocky Mountains. Duerksen *et al.* (1997) made recommendations for protecting habitat at both local and regional scales. They assumed that regional land-use changes are the result of many individual decisions made at local scales. Residential development influences wildlife at both local and regional scales – at the regional scale, species distributions and persistence are affected, while at the local scale factors such as the behaviour and reproduction of individual organisms is affected. Prevention and mitigation of these impacts can therefore be approached at both scales. Local management actions included: (1) preserved 'buffer' requirements for new developments; (2) use of native species in site plantings; (3) control of pets such as cats that can be harmful to local wildlife; (4) minimising human contact with sensitive species (e.g. large predators, in this case); and (5) facilitating movement across built areas, e.g. using wildlife tunnels under roads. Regional management suggestions included: (1) conservation easements, where land on a development is donated for ecologically friendly use in perpetuity; (2) zoning to restrict level and type of development permitted; (3) more effective transportation planning to avoid unnecessary ecosystem fragmentation; (4) protecting rare landscape elements (e.g. wetlands and riparian zones); (5) identifying and protecting landscape corridors; and (6) maintaining important ecological processes where possible, such as fires and flooding. Regional strategies were considered most likely to be effective in exurban and non-urban areas where large natural areas may still exist, while local strategies were perhaps the best (or only) course of action within urban complexes. These recommendations were based on landscape ecology theory and what was known from other applied studies. They were implemented by the Colorado Natural Diversity Information Source (NDIS), which provided information on possible impacts to developers.

A decision-support system for ecologically sensitive planning was also developed for Summit County in Colorado, which was experiencing rapid urbanisation. This system was created following collaborations between geographers, ecologists, GIS analysers, lawyers, landowners, developers and planners, and essentially used existing data such as species distribution and vegetation cover/classification maps to determine relative habitat values for different areas. A GIS map of such values could then be used to determine potential impacts of urbanisation scenarios in different locations, to help guide which scenarios might preserve important regional habitat, and to allow an initial assessment of potential impact for specific developments (Duerksen *et al.*, 1997). This is an excellent early example of a regional decision support system, and highlights some of the principles that later models (e.g. Mörtberg *et al.*, 2007) have followed.

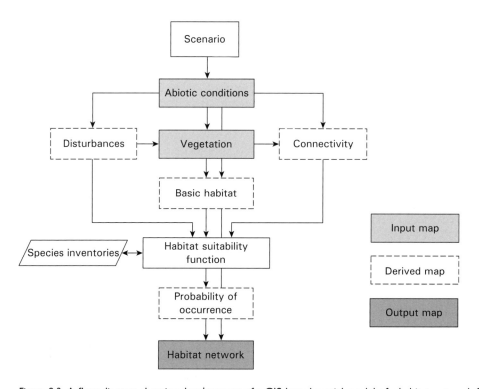

Figure 8.2 A flow diagram showing development of a GIS-based spatial model of a habitat network for a selected species. Once a basic habitat model for a defined area is created, species inventories inform the suitability of the available habitat for the species, to allow prediction of probability of occurrence across its spatial extent. This may then be used to create a complete habitat network map, to inform likely impacts of urban development or ecological improvement. (Reproduced from Mörtberg *et al.*, 2007, with permission.)

(*Tetrao urogallus*) and hazel grouse (*Bonasia bonasia*), both of which require old-growth forest and wetland habitat and therefore address target (1), and willow tit (*Parus montanus*), which has intermediate habitat requirements and was therefore more related to target (2). The principle followed was that as these species are sensitive to such habitat loss, they would reflect the worst-case situation for biodiversity loss if forests decline due to the urbanisation scenarios. Simple habitat maps were then created using GIS, based on land use, wetland presence, vegetation classifications, etc., and then used to model probability of occurrence of the species in different areas based on habitat suitability. These were then compared to habitat suitability maps predicting the three urbanisation scenarios. Results indicated that the 'densification' scenario in which the complex was kept compact was the least disruptive to the focal species, while the diffuse expansion model was most disruptive. Expansion along transportation corridors was as disruptive as diffuse expansion for capercaillie and willow tit, but not hazel grouse. This exercise demonstrated the most appropriate form of housing development for the region, at least based on the selected targets of forest preservation, and is a good model for future efforts in ecologically sensitive urbanisation.

Perhaps the most ecologically beneficial form of urban planning and design is the preservation, creation and careful management of green space. Some key ecological principles are also being applied in this case, as discussed below.

8.3 Green space planning and design

Ecologically based planning and design of green space began to emerge as a forma-lised concept in the Netherlands in the 1960s, though related ideas had developed since the early twentieth century (Jongman *et al.*, 2004; Adams, 2005). Fundamental ecological concepts such as succession (natural change in ecosystem composition; see Section 6.5), landscape connectivity (Sections 2.5.1 and 8.4), and the preferential use of native species rather than non-natives (Section 6.3.1) began to be explicitly incor-porated into urban planning and management of green space. These principles were then adopted into planning and design in other parts of the world, particularly the UK and USA (Adams, 2005). Spatial integration of green space into green networks to maximise connectivity is discussed further below, while other aspects of green space planning and design are briefly considered here.

8.3.1 From landscaping to wilderness

Many planned green spaces, such as urban parks or gardens, were 'landscaped', or carefully designed to create a semblance of rural countryside or a domesticated 'wild-erness' within the urban complex ('*rus in urbe*', or 'the countryside within the city'). The history of garden design and management is one of demonstrating control over nature and the creation of personal idylls (Turner, 2005). This is somewhat under-standable: as a species, we have spent thousands of years cultivating a 'controlled' nature for resource acquisition and to reduce the perceived threat that 'wilderness' can represent (e.g. Jorgensen and Tylecote, 2007; Francis, 2009b). Distancing from 'wilderness' and its associated hardships is still a major driver of urbanisation in some developing countries, where it is not romanticised in the Western tradition.

The recognition that natural regeneration of an urban green space via succession leads to more complex habitat and vegetation structure, and therefore more niches for other species, has led to an agenda of 'making space' for wilderness within green spaces. These include wildlife gardening, wherein wildlife is encouraged to use gardens by increasing habitat availability (e.g. unmanaged patches, ponds, nest boxes, dead wood), feeding wildlife, and reducing use of herbicides and pesticides. Some examples of techniques are given in Gaston *et al.* (2005b) and Thompson (2006). Urban parks are now more likely to contain unkempt and relatively unmanaged areas that may sup-port species, or to retain elements of pre-urban ecosystems such as natural woodland patches. Succession may be managed or unmanaged – allowing an area to develop naturally will generally create a 'wilder' environment and maintain heterogeneity (Section 8.3.2), though managed succession may involve the intentional maintenance of particular habitat mixtures and vegetation structures. Unmanaged succession is more likely to take place in private gardens, brownfield sites or areas of woodland where recreation and human access is less important, while managed succession will be the norm in parks, gardens and woodlands that have important recreational or aesthetic facility. Recognition that natural ecosystem dynamics are both important and desirable is a significant step forward.

However, urban wilderness may still be viewed as complex, chaotic and 'untidy' and may be associated in the minds of citizens with 'neglect'. This may influence engagement with green space and wildlife and can be observed in, for example, negative perceptions of urban woodlands or brownfield sites (Jorgensen and Tylecote, 2007), and the unwillingness of citizens to participate in conservation activities such as habitat enhancement in domestic gardens if they felt disapproval from neighbours would be forthcoming (Gaston *et al.*, 2005a). Indeed, 'wilderness' re-creation is most likely to be supported by those who do not live in the vicinity (Jorgensen and Tylecote, 2007). This is a further consideration relating to planning and design of green space, in particular those areas that are multifunctional and aimed at providing recreational services (see Section 8.3.4).

8.3.2 Habitat heterogeneity

The importance of habitat heterogeneity in creating abundant niches and allowing for the development of diverse species assemblages has been shown in Chapter 2. The strategic creation of areas of different vegetation type, height, and density, standing and fallen dead wood, vegetated and non-vegetated waterbodies and topographical variation that may create a range of microhabitats, may help to make individual green spaces more supportive of a range of species (e.g. Gaston *et al.*, 2005b; Hodgkison *et al.*, 2007). Though contradictory to the usual trend of managed neatness, this principle is beginning to be incorporated into design and management. Li *et al.* (2005), for example, recommended that natural forest structures (canopy, understory, shrub and herbaceous layers) be maintained in urban parks of Beijing, as part of a region-wide plan for urban greening. Likewise, the Olympic Park development in the UK utilised the redevelopment opportunity to create a variety of ecosystem types, to both support established species and encourage colonisation of new species (Box 7.1). There is, however, something of a trade-off between heterogeneity and habitat patch size – a more varied habitat by its nature contains smaller areas of particular types of habitat, and so some species may be disadvantaged if they need a certain area of habitat to persist.

8.3.3 Use of native species and control of non-natives

Historically, design and management of many green spaces (particularly parks and gardens) involved the planting of non-native species as they were of horticultural interest. While this remains common in many cases, increasing awareness of the risks of planting non-natives that may become invasive has led to an increasing tendency to plant only native species, ideally of local provenance to ensure establishment success, and the active control of non-natives (Li *et al.*, 2005; Goddard *et al.*, 2010). Non-native exclusion and control is more likely to result when species may represent a potential health risk, from new allergies resulting from exotic plant pollen to skin lesions from toxic sap, as in the case of giant hogweed (*Heracleum mantegazzianum*) (e.g. Cariñanos and Casares-Porcel, 2011; Pergl *et al.*, 2012). Despite some of the problems in identifying and controlling invasive species in both public and private green space, they may represent an important focus for control of non-native and invasive species for the wider region (e.g. Ishii *et al.*, 2010; Section 6.3.1).

8.3.4 Social inclusion and inequities of green space

Green space in particular has important societal value, alongside its ecological importance. Urban planning and design therefore needs to incorporate policy relating to green space provision and access, which is usually considered in terms of distance from residences, proportion of land cover, or area per capita. It can be important to ensure that management does not discourage use of green space (unless desired), and to encourage inclusivity, particularly for successful environmental stewardship. There is abundant evidence that socio-economically disadvantaged communities have reduced access to green space in many urban areas, which can lead to inequity in green space provision and use (Section 3.9.1). This is one factor behind an increase in urban greening, as many governments aim to increase green space access and environmental justice across urban regions. Cultural differences also emerge in relation to green space. In some cases socio-economic factors are linked to culture, with, for example, minority groups tending to have lower socio-economic status in many urban regions, in turn relating to the inability to choose desirable areas to live in, or access green spaces using private transport. However, some differences are purely cultural. It has been demonstrated that ethnic minorities often perceive, appreciate or use green spaces in different ways to the dominant (and generally decision-making) ethnic groups. Byrne (2012) found that the some members of the Latino community felt excluded from the urban Santa Monica Mountains National Recreation Area of Los Angeles due to perceived 'white' typology of the park, and uncertainty of social norms for such locations, leading to a lower proportion of Latinos visiting the park compared to white ethnicities, despite their living in close proximity.

In many Western urban regions, ethnic minorities such as Afro-American, Latin-American and Islamic groups have been found to prefer green space that is well-managed and clearly artificial, rather than a simulated 'wilderness' (e.g. Virden and Walker, 1999; Buijs *et al.*, 2009). This is likely linked to cultural perceptions regarding nature and humanity's relationship with it, but also the ways in which different groupings use the green space. In a study of Turkish and Moroccan immigrants in the Netherlands, Buijs *et al.* (2009) demonstrated that immigrants tended to use the green space for family gatherings (which is facilitated more by well-planned, maintained, open spaces) rather than for more solitary pursuits such as walking, hiking or biking (as favoured by the native citizens and which is facilitated by simulated wilderness). It is not always apparent whether this is an 'immigrant' cultural effect or a genuine cultural preference – for example, Jim and Chen (2006) observed that native Chinese communities prefer managed space but tend to use urban green spaces for solitary activities (exercise, contemplation) rather than large family gatherings; but also that there is little research into Chinese immigrant use in Western urban regions.

These are generalisations to an extent: perceptions and use will vary depending on personal preference and residency time (Johnson *et al.*, 2004) among other things, and in many cases the data collected do not classify communities in sufficient detail. It is nevertheless clear that in general, immigrants do tend to utilise green space within urban complexes rather than outside, have reduced access to green space, and prefer more managed as opposed to 'wild' spaces. This may be particularly important in a planning context given that the main population growth in many urban regions is due to immigration rather than births, and given that cultural groups tend to cluster (Kinzig *et al.*, 2005; Forman, 2008).

8.4 Spatial planning of green networks

The spatial configuration of green space is important for ensuring connectivity and effective functioning (see Chapter 2). The concept of 'green networks', has recently grown notably in popularity as research into landscape ecology and diversity has increased, and because the concept – or at least the visualisation – of landscape connectivity is one that a wide range of stakeholders can potentially understand (Vimal *et al.*, 2011). Several different terms relating to green networks are used in the urban context (Table 8.1). A green network may be conceived as patches of green space connected by corridors running through the intervening matrix. In network theory, the patches would be termed 'nodes', while the corridors would be 'linkages'. The patches

Table 8.1 Definitions of commonly used terms associated with urban green space planning and design

Term	Definition
Green network or ecological network	A series of connected green spaces within an urban region, usually composed of patches linked by corridors. The principle of connectivity is fundamental to a green network, as a 'network' would not exist without it. The aim of the network is usually to allow dispersal of organisms through the fragmented urban landscape, in order to maintain metapopulations and ensure long-term survival (persistence) of species
Green web	A synonym for 'green network' (Turner, 2006); also used to refer to environmental communications networks in social science literature
Green grid	A synonym for 'green network', mainly used in a UK context
Green belt	A ring of green space encasing an urban complex, usually with some form of legal protection against development and intended to prevent urban expansion
Greenway	A predominantly American term but widely applied, used to refer to linear green space alongside a natural or artificial corridor (e.g. Jongman *et al.*, 2004). A green network may be constructed in whole (rarely) or in part (more frequently) from greenways. In some cases 'greenway' is used as a synonym for 'green corridor'
Green corridor	Generally synonymous with 'greenway', but does not necessarily imply a linear structure. May simply be a collection of green space that allows species dispersal through an urban landscape, usually between distinct patches of green space
Parkway	Generally an American term, denoting linear tracts of land set aside for recreation, usually alongside a road and associated with recreational driving. Somewhat synonymous with 'greenway' but usually suggests wider corridors and limited use of the vegetated areas (see Erickson, 2004). Often extends from urban to non-urban areas

(nodes) essentially represent important habitat, for example locations where reproduction may take place or where 'interior' species may establish, while corridors
(linkages) may allow dispersal between patches, though this is a simplistic interpretation. Network corridors are often conceptualised as relatively linear, and some may be
(e.g. roadside greenways), though in reality they may also be stepping stones or patches
rather than a contiguous corridor – the key aspect is that they allow dispersal of organisms and persistence of populations within the network.

Network corridors are also often designed to be multifunctional, for example
improving recreation or aesthetics or reducing flooding impacts. This can ensure
greater sustainability and resilience of the urban ecosystem. They may, however, also
present some disadvantages; for example, a connected network may increase risk of
spread of disease or invasive species through the urban region. Examples of green network designs proposed for several urban regions are shown in Figure 8.3, though in
many cases designs remain to be implemented (e.g. Erickson, 2004). Networks are not
usually planned or designed from scratch but are rather a way of conceptualising the
green space that is found within an urban region, and for increasing its function, for
example by extending or widening corridors or increasing ecological quality in parts
of the network. Several broad principles may be adopted when considering the planning and design of green networks, which stem from the patch–corridor–matrix
model (see Box 2.1) and elements of network or graph theory.

8.4.1 Principles of green network planning and design

The ecological principles of the planning and design of green networks relate to established landscape ecology theory on ecological networks in the natural environment
(Forman, 2008). Essentially, they aim to maximise the quality of the network by making the individual patches and corridors function well, and by maximising connectivity through the network. The key principles are:

1   *Identification of natural or existing networks.* Ecological networks occur in nat-
    ure, and consequently quantifying those present prior to urbanisation can ensure
    that appropriate conditions are maintained. Likewise, urbanised areas will still
    maintain some distribution of green space that may function as a network or has
    the potential to do so, for example patches and strips of remnant woodland, park-
    land or river and canal systems. The key initial process in spatial planning is to
    determine core areas of biodiversity, 'buffer' areas (those that should ideally be
    protected from further encroachment/impact, or supported by further greening
    efforts) and both existing and potential connecting corridors (Bryant, 2006). This
    then forms a basis for further planning and design.
2   *Maximising patch and corridor size.* Following the species–area relationship (see
    Box 2.4), the network is likely to function best if patches and corridors are of the
    largest size/width possible, as this will increase their capacity to maintain species
    populations and support 'interior' species. Even if only a few patches or sections
    of corridor are large, these may act as source ecosystems within the network.
    Forman (1996) notes how the functions of corridors in particular (as habitats,
    conduits, filters, sources and sinks) are often enhanced with wider corridors. It is
    possible to try to ensure large patch sizes during urbanisation by safeguarding

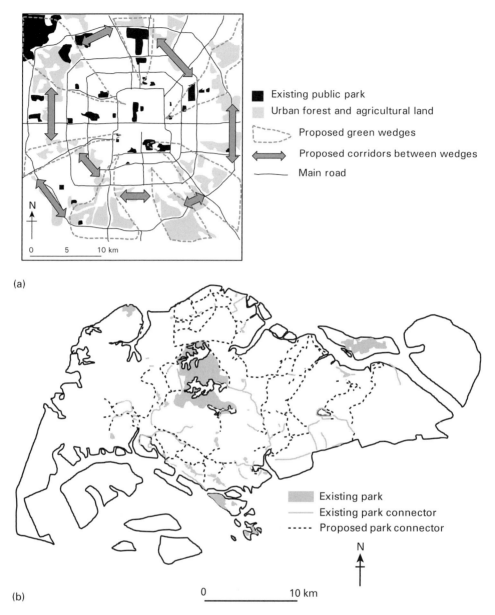

Figure 8.3 Examples of green network plans for four Asian urban regions. (a) Proposed green network for Beijing, based on existing areas of green space and suggested placement of green wedges and corridors. Within green wedges, removal or greening of residential areas was proposed to increase proportion of green space. (Redrawn from Li *et al.*, 2005, with permission.) (b) Proposed green network for Singapore, using multifunctional greenways to connect existing park areas. (Redrawn from Tan, 2006, with permission.) (c) Proposed green network for Xiamen Island, based on network analysis to determine the best configuration of greenways to connect existing green space. This diagram demonstrates the most feasible scenario. (Redrawn from Zhang and Wang, 2006, with permission.) (d) Proposed green network for Hanoi, Vietnam, highlighting roadside greenways to be constructed by 2020 as part of an overall green space planning strategy. (Redrawn from Uy and Nakagoshi, 2008, with permission.)

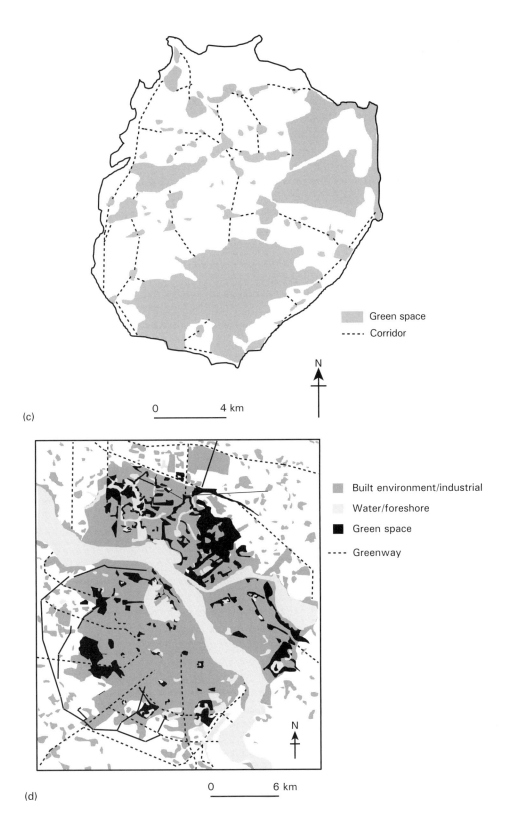

Green space
- - - Corridor

N

0   4 km

(c)

Built environment/industrial
Water/foreshore
Green space
- - - Greenway

N

0   6 km

(d)

these, and to expand patches by complementary plantings in adjacent areas (e.g. by increasing the provision of wildlife gardens near to a woodland patch) or by replacing urban land cover (e.g. housing) with green space (e.g. Li *et al.*, 2005). As it can be difficult to determine appropriate sizes and thresholds for patch and corridor functions, often measurements are based on key species (ideally those that have an important role in ecosystem function, but sometimes those of conservation concern), or by working within the practical limitations of the urban landscape.

3   *Increasing patch and corridor number and density.* Increasing number and density of patches and corridors maximises the cover of green space within a given area and also decreases distance between individual patches, potentially increasing both population size and dispersal. At sufficient densities negative influences of the urban matrix may also be reduced. Such increases in green space may be difficult in urbanised areas due to limited available land, and is most likely to focus on small, fine-scale additions of green space to the network, for example brownfield sites, gardens and roadside verges.

4   *Increasing habitat quality.* Habitat quality may be reflected in patch and corridor size, increases in heterogeneity and reduction of non-native species, as noted in Section 8.3, but may also include ensuring that all habitat characteristics required by the focal species are present (e.g. nesting sites, prey species, vegetation or soil types, suitable water quality) and that obstacles such as low-quality areas, gaps in corridors or other physical barriers are absent or mitigated.

5   *Connecting patches with corridors.* A patch connected to a corridor is generally considered to function better than one that is not connected, though in many cases the evidence for this is inconclusive (see Section 2.5.3; Angold *et al.*, 2006). Certainly proximity to other green space is important in facilitating movement of organisms between patches, though for many species a series of patches acting as 'stepping stones' is just as effective (if not more so) than linear corridors. For some species the contiguous habitat represented by a corridor may be important. Forman (1996) suggests that straight, wide corridors are the most effective for guiding species movement and facilitating dispersal through a network, but there is little empirical evidence to support this. Nevertheless, corridors remain a key aspect of green network planning and design, and are often the focus of such plans (Figure 8.3).

6   *Maximising circuitry where possible.* Alongside connectivity, networks generally work best when they maintain circuitry, which is the presence of 'loops' or 'circuits' in the system (Figure 8.4). This allows greater movement around the network, and essentially offers alternative routes for (for example) individuals to disperse or forage. A network with greater circuitry means that a disruption to the network, for example the creation of a barrier, corridor gap or area of poor habitat quality or high mortality, is not as severe as in a network with no or limited circuitry (Figure 8.4). Such a system may be considered to be more 'resilient' to network disturbances (see Box 3.1). Consideration of network connectivity and circuitry often go hand in hand, and some green networks have included circuitry metrics in their planning (Box 8.2). This does not apply to all networks, however – river networks, for example, have no real circuitry (though they may be highly connected). This is one reason why the interruption of the linkages within such

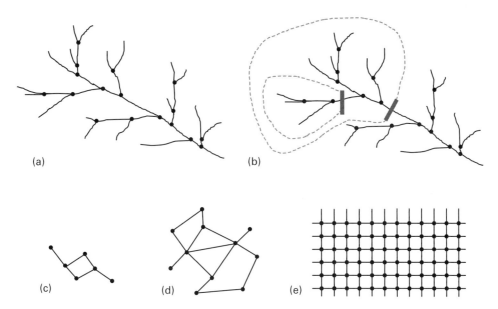

Figure 8.4 Examples of networks and circuitry. (a) represents a branching network, in this case a dendritic network such as that observed in a river system. Although the system is connected, there is no circuitry as no loops are present. Consequently, an interruption to the network may isolate substantial proportions of the network. This is demonstrated in (b), which illustrates how dam construction may isolate sections (grey dashed lines) of the network, restricting flows of some materials to and from these sections. Circuit networks are shown in (c) to (e). (c) shows a weak circuit network with limited circuitry (circuitry may be measured by dividing the number of circuits or 'loops' present by the number that could be made possible were extra linkages put in place; Forman, 1996). (d) shows a larger network with more circuitry, while (e) shows a grid network, with very high circuitry. This is why highly planned transport networks may display grid-type layouts.

systems can be particularly problematic, and have significant repercussions both upstream and downstream. Terrestrial green networks have greater capacity for circuitry, and this is essentially one of the aims of proposed corridor creation in many cases.

7 *Allowing natural dynamics.* Planning of green networks mainly considers the spatial aspects in detail but not so much the temporal aspect of ecosystem change; Ramalho and Hobbs (2012) recognise this as a key aspect of successful future planning and design, within a 'Dynamic Urban Framework' that incorporates environmental change over time. Patch and corridor configuration and quality will change over time as succession takes place and due to more stochastic change as species and populations shift in location and abundance, and environmental conditions alter. Ideally, although all green space will probably need management at some level, this would not be too proscriptive and would allow natural seral dynamics to take place (Kattwinkel *et al.*, 2011). This may be particularly important when considering the long-term future of green networks, for example under urban densification or climate-change scenarios.

Box 8.2 A green network for Xiamen Island

Xiamen Island is the centre of Xiamen, one of China's fastest growing urban regions. Zhang and Wang (2006) evaluated the potential for creating a diverse and sustainable urban green network within the region, using a combination of landscape metrics and network analysis. They began by using satellite images to map and classify land use, including urban, agricultural, water, coastal wetlands, natural green space (e.g. woodland) and cultivated green space (e.g. parks, gardens, golf courses). Landscape metrics were then calculated based on the GIS maps created, with the key metrics including measures of patch area, density, shape and diversity; edge density; isolation and connectivity.

This data was then linked to network analysis, which used measurements of corridor number, length and density to calculate network characteristics such as circuitry and connectivity. Using this methodology it was then possible to consider varying green network augmentation plans and how they would affect the landscape metrics measured. A plan had already been proposed for the region in 2002, and the analysis demonstrated that this plan would change the cultivated greenway area from 1491 ha to 3434 ha, patch size from 3.8 ha to 9.5 ha, patch density from 2.98 patches/ha to 2.74 patches/ha and edge density from 29.7 m/ha to 45.4 m/ha. In total this means an increase in both area and structural connectivity (i.e. a decrease in fragmentation) of green space for the region. The natural green space would remain the same.

A network map was then created to suggest further greenway developments, to provide greater support for the augmentation plan (as shown in Figure 8.3). This network provided optimal circuitry and connectivity, based on which corridors could feasibly be created, and demonstrating further increases in the landscape metrics measured, reinforcing a potential increase in green space area and connectivity. This is an excellent example of ecological urban landscape planning based on both landscape ecology and network analysis to maximise potential use of urban space.

8   *Connecting beyond the region.* Green networks ideally consider not just the urban complex or the urban region, but also non-urban areas that may have high biodiversity and maintain source populations that may support the urban network (Turner, 2006). The urban green network should be thought of as part of a wider network – one that may potentially be ecologically poor but which nevertheless is unique and of notable importance to urban residents. Though difficult to coordinate, a wider network of green space running within and between urban and non-urban regions is perhaps the most likely to be both biodiverse and resilient.

8.4.2 Determining connectivity

Although it is widely recognised that a more connected landscape will function better ecologically, it can be difficult to determine exactly what 'connectivity' means for a given species or process (see Section 2.5.3). This can be particularly important for urban planning; the effectiveness of green networks is limited if they are not actually functioning as a network, i.e. facilitating the spatial movement of species or materials (e.g. water). This will in many cases depend upon what the network was intended to achieve. A network planned for recreation does not need to consider species movements, but a network aimed at sustaining metapopulations of particular species, or maintaining biodiversity in general, does. Baguette and Van Dyck (2007) argue that appreciating the difference between structural and functional connectivity is crucial if green network implementation is to be effective. Most landscape or network metrics measure structural connectivity (Kupfer, 2012). However, this must be linked to reliable information on species movement or dispersal between network components (particular corridors and patches, for example) to determine whether the network is functioning in an appropriate way. This is a particularly complex problem because:

1 Different species will interact with networks at different spatial scales (resolution), meaning that a network will display different patterns and trends for individual species (different species 'perceive' their immediate habitat differently).
2 Differential dispersal will be observed within particular populations and individuals within a species, according to different environmental pressures, meaning that populations of the same species may disperse differently within different sections of a network, even though the perceived 'quality' of that section of the network may be the same.
3 Both dispersal and connectivity will vary as the landscape changes, for example as species populations change in size (a good year for reproduction creating more dispersal of juveniles due to population pressure, for example) or as seral changes alter the environmental conditions of network corridors and patches.
4 Dispersal is not uniform spatially but may be considered as three stages: emigration (moving out of an area); transfer (movement across an area without long-term settlement); and immigration (settling in a new area) (see Baguette and Van Dyck, 2007). Networks are designed to facilitate dispersal in general, but each of the stages may operate based on different parameters (corridor width, patch or edge quality, vegetation characteristics, organism vagility and so on), meaning that different network sections may only facilitate certain aspects of dispersal.

This does not mean that green networks are not effective, however, or that they should not be a staple of ecological urban planning. But it is important to be cautious when assessing network effectiveness, and oversimplifying decisions based on structural metrics. Models may be developed to establish how individual species or species groups may utilise spatial configurations in the urban landscape, but these may often be problematic (Baguette and Van Dyck, 2007). There is still a long way to go in determining the ecological effectiveness of green networks. As a result of this, further efforts to 'green' the urban matrix, at least in support of any networks that may exist, are certainly a step forward.

8.5 Ecological and environmental engineering of urban ecosystems

There is increasing interest in the ecological and environmental engineering of urban ecosystems (see Section 7.1 for the distinction between ecological and environmental engineering). This essentially means the enhancement of green space or the built environment with some form of engineering, and is usually multifunctional, helping to mitigate an environmental impact as well as providing habitat. In some cases, green spaces are directly enhanced, for example with the construction of wetlands as part of a sustainable urban drainage system (SUDS), which may provide a water supply while at the same time reducing flood risk and pollution, and providing urban greenery. In other examples, buildings are engineered directly to allow species to utilise them as habitat, while enhancing urban drainage and climate amelioration. In this section, two examples of ecological and environmental engineering techniques that are being incorporated into urban planning and design are presented: SUDS and the ecological engineering of built surfaces.

8.5.1 Sustainable urban drainage systems

SUDS are incorporated into urban planning and design to more sustainably and ecologically manage urban hydrological cycles and water use. They usually consist of porous or vegetated building materials, vegetated 'buffer' zones and biofiltration and storage systems that together help to reduce urban runoff and associated problems (see Section 3.6), and allow the harvesting, treatment and storage of urban rainfall or (in some cases) floodwater. The components together make up a 'treatment train' whereby rainfall or runoff and any associated contaminants is retained and sometimes filtered at different stages, from initial rainfall (retained by greened buildings, soils or porous materials) through buffers (e.g. vegetated areas along infrastructure or impermeable surfaces that help to retain water and pollutants) through to specifically designed biofiltration systems that improve water quality before releasing or storing the water. 'SUDS' is the term generally used in the UK, but such systems are also referred to as water-sensitive urban design (WSUD) in Australia and low-impact development (LID) in the USA (Coutts et al., in press). As with green networks, they may be conceptualised at the landscape scale but are more commonly implemented at the local scale. Often planning policy guidelines specify that some form of sustainable drainage system or stormwater harvesting be put in place if an urban development is to occur, particularly if a substantial area of impervious surface is to be created (Coutts et al., in press).

Their main purpose is to retain water within the urban complex so that the negative effects of increased surface runoff are reduced, and to allow treatment of the water to make it more useful for urban water requirements (see Section 3.6). They usually store excess runoff and in some cases (e.g. in wetlands constructed adjacent to rivers) floodwaters, at least temporarily. This therefore helps to reduce heavy inflow of water to conventional drainage systems or rivers, and may also allow the water to be stored or used within the region. Such water stores may be particularly important in areas where water supplies are limited (Coutts et al., in press). Retained water may, for example, be used for irrigation purposes, particularly if it has been stored within a constructed wetland or other biofiltration system designed to remove pollutants, including sediments, heavy metals or nutrients (Hogan and Walbridge, 2007).

Constructed wetlands are usually areas of vegetation planted on a porous substrate into which water is directed. The plants and the micro-organisms they support help to break down and remove pollutants in the water, thereby cleansing it for other uses (e.g. irrigation). They may be further divided into: (1) subsurface flow, wherein water is directed through the wetland substrate (usually composed of relatively permeable sand or gravel); and (2) surface flow, where water is conducted across the soil and through the aboveground vegetation, and which may be based on a variety of soil types. Specific wetland designs vary according to area available, the type and volume of water to be treated, substrate and species planted, and local climate (e.g. Hijosa-Valsero *et al.*, 2012). Often, reed beds are planted, with *Typha* spp. and *Phragmites* spp. being widely used. Constructed wetlands have been shown to be reasonably effective in pollutant removal over both short (1–3 years) and longer (5–7 years) timescales, with different design configurations (e.g. species selection) being more effective for certain types of pollutants (e.g. Mustafa *et al.*, 2009; Hijosa-Valsero *et al.*, 2012). Urban constructed wetlands are often found in wastewater treatment plants, riparian zones, on golf courses, and in other forms of green space. The cleansed water may be stored in tanks or reservoirs, or may require further treatment to become of potable quality and consequently used directly for human consumption or to recharge depleted aquifers (Page *et al.*, 2010). They may also represent important habitat for wetland species, particularly in urban areas where such ecosystems and under-represented (Hamer *et al.*, 2012).

More recently, SUDS have been considered alongside green space in general as a form of engineering that may ameliorate urban climate conditions. Coutts *et al.* (in press) have suggested that SUDS can be strategically designed and positioned to support green spaces in lowering urban temperatures and improving human comfort. Alongside creation at regular intervals throughout the built environment to help maintain sustainable urban hydrology and water use, they may be targeted specifically at areas of high temperature, such as where vegetation is scarce. Such coordination remains relatively unexplored at the present time, and potential success depends on the wider climatic region, the characteristics of the urban region in question and individual SUDS designs.

8.5.2 Ecologically engineered built surface materials

The active greening of buildings using living roof and wall installations may provide habitat while helping to ameliorate surface runoff, flash-flooding, atmospheric urban pollution and heat-island effects (e.g. Obernforfer *et al.*, 2007; Escobedo *et al.*, 2011). The effectiveness of such structures over different timescales is still being evaluated, though it seems that the potential may exist to improve both biodiversity and urban environmental quality in different climates and development scenarios (e.g. Francis and Lorimer, 2011). Living roofs and walls have been briefly described in Chapter 5. They are now more specifically being incorporated into urban planning and design policy, with many countries including them within options to meet environmental sustainability targets, or specifically requiring their construction in certain planning zones or developmental scenarios, e.g. for new governmental or public buildings (Carter and Fowler, 2008).

Surface greening may also be performed along road infrastructure. Cao *et al.* (2010), for example, demonstrated that air bricks (bricks manufactured with sizeable holes or slits) filled with soil can support abundant vegetation growth while maintaining sufficient strength and integrity to act as retaining walls (i.e. preventing slope slippage) along roadsides. Temporary structures such as construction site screens may also be greened, though this is usually purely for aesthetic purposes (Francis, 2011).

Other forms of built surface engineering are also being developed. Many urban regions are coastal and have extensive hard-engineered flood defence structures that degrade estuarine and coastal ecosystems. There has been a substantial body of work exploring the potential of ecologically engineered sea walls, focusing in particular on Sydney, Australia. Novel attempts to incorporate artificial habitats such as tidal pools or other forms of structural complexity in the wall structures have shown promise in increasing the diversity of marine and estuarine invertebrates (Chapman and Underwood, 2011). Browne and Chapman (2011) demonstrated that the addition of wall cavities and attached pots to seawalls in Sydney harbour more than doubled species found on the walls, including the colonisation of species that would not be found without such modifications.

Some techniques may have the potential to be revolutionary and dramatically change the appearance and function of urban buildings. The Institution of Mechanical Engineers (2009) have highlighted the potential for buildings to be covered in photo-bioreactor systems containing algae that photosynthesise and store carbon as they grow, a form of climate control that may be considered geo-engineering. The algae grown in these systems may also remove atmospheric pollutants, and potentially be used for biofuels or other applications. Although this form of technological design remains at the conceptual stage only, such innovations have the potential to be trans-formative within urban regions.

8.6 Chapter summary

1   Planning for urban ecology is a relatively recent development – although some vegetated ecosystems have been preserved or created in urban areas for millennia, this has been to improve the urban environment for humans rather than any parti-cular ecological motivations. It should be noted that urban planning and design vary between countries and regions, particularly in relation to ecology. Many of the examples discussed in this chapter are from developed countries with exten-sive urbanisation, which have been the focus for most urban ecology.

2   There is now a move towards ecologically sensitive urbanisation. This is often supported by legislation and policy and may involve both prevention and mitiga-tion of impacts. Usually this is performed at a local scale, though some initiatives have focused on landscape-scale ecologically sensitive planning.

3   Green space is increasingly being managed according to ecological principles, including greater incorporation of 'wilderness' characteristics, habitat heteroge-neity, non-native species control and greater social inclusion.

4   Spatial planning of networks at the landscape scale is becoming more popular, though may often remain at planning or design stages. Several principles have emerged from such spatial planning, including the importance of identifying nat-ural or existing networks, maximising patch and corridor sizes and densities, increasing habitat quality, and maximising network circuitry.

5   Ecological and environmental engineering techniques are also being developed. Examples include SUDS, greened buildings and infrastructure and built sur-faces that may support a range of species and/or provide other environmental benefits.

8.7 Discussion questions

1 Does your urban region maintain a green belt? What factors might have determined this? Has it been effective? How might you evaluate this?

2 Design a model for assessing impacts of urbanisation on a hypothetical species. What information would you need? Construct how the model/decision-making process might work (e.g. Figure 8.2).

3 How would you respond to increased urban 'wilderness'? What could be done to increase social inclusion and provide green space for everyone? How might people be better educated about the value of 'wilderness'?

4 Consider what the three main challenges for improving green networks may be. Why do many green networks not get implemented, or only partially completed? How useful is the green network concept?

5 In what ways are urban planning and design likely to differ between developed and developing regions? How might capacity for ecological urban planning be improved?

Chapter 9

The future of urban ecosystems

9.1 Introduction

The previous chapters have provided an overview of some key physical, social and ecological aspects of urban ecosystems. The goal of this work has been to provide a sound foundation for further exploration of urban ecosystems and urban ecology. However, the understanding of urban ecosystems and the socio-ecological patterns and processes that drive them remains in its early stages, and the story of our new human environment remains far from complete. Certainly, whatever decisions are made and whatever trajectory our species and civilisation continues along, urban ecosystems are likely to continue to grow, develop, be studied, be refined, reconstructed and reconstituted, and ultimately to feature heavily in our future. This concluding chapter considers the future of urban ecosystems and urban ecology, highlighting trends that are likely to continue into the coming decades and their potential significance for us and our environment.

9.2 Urban growth

Urban areas are sure to continue growing for decades at least. It is predicted that population growth will continue until a peak of between 8 and 10 billion people is reached, probably around the middle of the twenty-first century (Lutz and Samir, 2010), when a rough equilibrium will be established. The majority of this population growth is likely to occur in urban areas, particularly in Africa, Asia and South America (Figure 9.1; Lutz and Samir, 2010; UN-Habitat 2011). It is likely that 70 per cent of the global population will be urban by 2050, with regions such as North America, South America and Europe having well over 80 per cent urban populations, and Africa and Asia having over 60 per cent (UN-Habitat, 2011). In particular, mega-cities or 'meta-cities' (considered in different publications to be urban regions with greater than 5–20 million people, with 10 million being the most common threshold for megacity status) are likely to become more common, and there will also be an increase in 'mega-regions', as conurbations continue to expand and merge (UN-Habitat, 2011). Large mega-regions include Hong Kong–Shenzen–Guangzhou in China (population 120 million), Nagoya–Osaka–Kyoto–Kobe in Japan (predicted 60 million by 2015) and São Paulo to Rio de Janeiro in Brazil (43 million) (UN-Habitat, 2011). These vast urban regions and the overall increase in urban populations, particularly in developing countries, are likely to lead to increasing slum

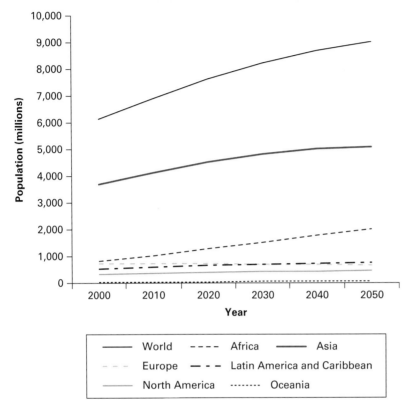

Figure 9.1 Predicted population growth from 2000 to 2050, both globally and for different regions. (Based on Lutz and Samir, 2010.)

populations and greater socio-economic inequity. This may be particularly the case in those regions experiencing immigration of refugees and displaced populations, whether this is from war and conflict or environmental degradation that has compromised their livelihoods. With unregulated intrastate conflict likely to increase in the coming decades (e.g. Metz, 2000), and increasing pressure on environmental resources and livelihoods from population rise and climate change, developing urban regions are likely to bear the brunt of such displacement. Globally, the number of people in slums is expected to grow by around six million people per year, reaching a total of around 889 million slum-dwellers by 2020. Though increases in living standards are helping to reduce the proportion of people living in slum-quality housing, this remains a substantial proportion of people living in slums (UN-Habitat, 2011).

Such urban development is likely to be categorised by rapid expansion, both from unplanned exurban slum developments by those in poverty, and by low-density residential expansion by the increasingly affluent. Rates of expansion (diffusion) are therefore likely to be relatively high until the middle of the twenty-first century, with densification (coalescence) occurring more slowly. This is likely to exacerbate the ecological impacts associated with such urbanisation, with further loss of habitat, ecosystem fragmentation, and degradation of environmental resources and ecosystem services.

This trend may be avoided or lessened by a decline in population growth and concentration in urban areas, but at the present time this seems unlikely. Fertility rates are generally declining due to increased education (particularly of females) and economic development in many parts of the world, and are now roughly at replacement levels globally (May, 2010). However, as the population has been growing substantially in recent decades, creating an abundance of younger generations, current fertility levels and increasing longevity will maintain this growth until the mid twenty-first century. Further attempts to encourage regional population control have been considered both unethical and unrealistic in some quarters; those regions that are most in need of control are often those where advances in education and economic development may be most difficult to achieve. Some researchers have suggested that rather than limiting growth, the focus should be on reducing per capita ecological footprints to allow greater numbers to coexist. Much urbanisation impact will depend on both how populations develop and whether we can limit impacts and increase sustainability. Based on this background, ecological management and mitigation of impacts relating to continued urbanisation seems the most realistic response over the coming decades. It will be a particular challenge to ensure that this response is significant within those regions experiencing most rapid urbanisation, and which may have the least resources to bring to bear.

9.3 Urban species and biodiversity

At the global scale, urbanisation and its associated loss of native species, in particular specialist species associated with lost or fragmented ecosystems, will continue to exacerbate the rapid rate of global species extinctions – a legacy that is already somewhat unavoidable (Rosenzweig, 2003). The common trend observed for localised decreases in diversity of some taxa but increases in others (e.g. plants) is likely to continue as urban regions expand (McKinney, 2008). Novel species assemblages will continue to emerge, as both native and non-native species are introduced to urban areas both intentionally and accidentally. The proportions of non-native species in assemblages are likely to continue to increase, at the expense of native species. Despite an increasing awareness of the threats of non-native species and screening and quarantine measures put in place to prevent introductions, the sheer scale of global movements of people and goods means that such introductions are somewhat unavoidable. This may be particularly true in developing regions that are most likely to experience continued rapid urbanisation (e.g. Asia, Africa and South America), due to a combination of increasing affluence (at least among some members of the population) that may increase global travel (e.g. for tourism) as well as movement of goods that may harbour non-native species, alongside a possibly reduced capacity to screen, quarantine or otherwise control non-native introduction and spread in such regions. Economic growth and increased globalisation drive both urbanisation and the movement of species (Hulme, 2009). Economic growth remains both a regional and global priority, and globalisation shows little chance of slowing down. Consequently, regulating the movement of non-native species between urban regions, while enabling continued development, remains a major challenge.

In most cases, datasets for urban species diversity are limited temporally, and it is not easy to reconstruct historical patterns from (for example) seed or pollen data, as urban soils and the evidence they contain is often destroyed or heavily modified by urbanisation (Pouyat *et al.*, 2010). Consequently, the baseline for much data is the present day, and it remains to be seen what trends may exist for species loss in urban ecosystems as current species and assemblages contend with the new urban environment (the extent of 'extinction debt' present in urban ecosystems). It will be several decades before a more detailed picture emerges for temporal changes and the future of urban biodiversity – for example, whether diversity levels will be maintained indefinitely by repeated colonisation of species from outside urban regions, as those within are lost.

Species will continue to respond and potentially adapt to urban ecosystems in interesting and surprising ways (Francis and Chadwick, 2012). As 'natural' species and ecosystems are lost, the growth of novel ecosystems and assemblages will create new selection pressures that over the long term may see urban ecosystems being crucibles of speciation, at least with regards to those species capable of persisting within such systems. In an essay on the future of nature over the next 100,000 years, Marshall (2012) has suggested that future ecosystems will be simpler than present, and that encouraging the development of new species, assemblages and ecosystems may be the most realistic and achievable way of coping with the biodiversity crisis long term. Within such a context, expanding and long-term novel ecosystems, such as urban ecosystems, may be locations where new 'founder species' may emerge. This is currently speculation of course, but the possibility indicates that more work is needed into urban ecosystems, their species, and their adaptations.

9.4 Urban sustainability

Urban sustainability is a complicated concept and a difficult objective for urban ecosystems, but will be important for the evolution of our human environment. It is important to remember that urban regions have particular societal functions, and that these must be satisfied alongside any ecological improvements. Urban ecological sustainability may broadly be considered the maintenance or improvement of ecological quality (in terms of ecosystem functioning and services) alongside economic and social development – i.e. the latter should not be detrimentally affected by ecological sustainability.

As noted in Chapter 3, urban ecosystems are maintained by heavy investment of external resources, both for direct resource use and for the processing of urban outputs. It is unrealistic to consider that an urban area may be entirely self-sustaining – the very things that define an urban region (abundant built environment and dense populations) mean that essentials such as food, water and fuel will not be obtainable within the land occupied by the urban region, despite technological advances such as SUDS or greening of the built environment, wider use of alternative energy sources or allocation of unused land for food production. This does not mean that such efforts are fruitless of course, and they may represent an important step towards reducing the ecological footprint of an urban region. Grewal and Grewal (2012) examined thee different scenarios for utilising available land and roof space for food production in Cleveland, Ohio, finding that in the most beneficial scenario (80 per cent of vacant land, 9 per cent of every occupied residential lot and 62 per cent of every industrial and commercial roof space), 7.3 per cent of food and beverage expenditure could be supplied. Even such schemes, if well coordinated and enacted effectively, will only make modest reductions of the urban ecological footprint.

Nevertheless, a move towards ecological sustainability may be achieved by: (1) reducing the urban ecological footprint via a combination of more energy-efficient building designs, use of SUDS, provision and sensitive management of green space, urban farming etc.; and (2) utilising urban ecosystems more effectively, in particular changing education and attitudes to urban biodiversity and recognising the importance and value of green space for ecosystem services. Given the limitations of preserving or creating green space in some cases, the possibilities of reconciliation ecology (see Section 7.4) should be considered in earnest. Such efforts may be particularly useful in developing countries, where high levels of biodiversity are often combined with increasing urbanisation – sustainability measures may have potentially substantial benefits in such areas, though this not where the majority of the scientific research is focused.

Urban metabolism models, whether based on energy equivalents or mass fluxes, may be effective ways to rigorously quantify the efficacy of the techniques utilised. There is an increasing focus on the development of indicators for the many facets of sustainability, and increasing attempts to get urban sustainability recognised as a major issue on the world stage (e.g. Boyko *et al.*, 2012). Success with the urban sustainability agenda will depend to some extent on both global and regional urbanisation trends, pressure on populations and livelihoods, economic cycles and the emergence of new technologies that may help to regulate ecological footprints. Realistic expectations need to be maintained – urban ecological sustainability may not be a major priority for many governments given consistent pressure to accommodate populations and maintain economic growth. Unfortunately, this is part of a wider trend for very slow societal responses to environmental crises (Mooney, 2010).

9.5 Urban conservation and planning

Despite some rapid advances, conservation science can be a time-consuming and iterative process, with time needed for concepts and hypotheses to be formulated, evidence collected, and then for guidance to filter through the relevant channels to inform conservation action, including ecological urban planning and design. Landscape ecology theory is only now becoming more common in urban planning, and often remains at theoretical stages. Implementation of landscape ecology principles is likely to increase, but may not occur as quickly as needed given projected rates of urbanisation. Despite being in early stages, ecologically sensitive urban conservation, planning and design tools are the means by which urban ecosystems may maintain their ecological quality and develop more sustainably. There is evidence that densification of the urban complex, preserving green space where possible, is the most ecologically sound planning option (Jabareen, 2006; Catalan *et al.*, 2008). Despite this, urban expansion is most generally predicted; where this is likely to occur, the landscape-scale holistic planning of preserved green networks should be adopted to increase the probability of biodiversity and ecosystem services being maintained, as well as providing important green space for citizens. This needs to be married to ecologically sensitive design in the built environment, so that green spaces and networks are not compromised by impacts from the surrounding matrix, such as stormwater runoff, localised pollution or spread of invasive species (Hostetler *et al.*, 2011). This may be best achieved by incorporating a systems science framework to planning, recognising that individual ecosystems within the urban region are part of the wider urban ecosystem. This principle needs to be more fundamentally applied to urban regions (Forman, 2008; Hostetler *et al.*, 2011).

Related to this is the explicit incorporation of the temporal dimension in conservation and planning; although this is an important element of ecological studies in general, there are relatively few studies that consider temporal change in urban ecology (e.g. Ramalho and Hobbs, 2012). As part of their 'Dynamic Urban Framework' (Figure 9.2) Ramalho and Hobbs (2012) explicitly incorporate a temporal aspect, noting that ecological patterns and processes can only be properly understood in the context of land use legacies, patch dynamics, succession and time lags of impacts (e.g. population or species extirpations) in such rapidly changing ecosystems. This is important not just for understanding patterns and processes but for proactive planning and design, and is likely to feature notably in future spatial planning.

Under both expansion and densification scenarios, greater consideration should be given to the application of reconciliation ecology via ecological engineering, for example the development of greened buildings and infrastructure. Although the evidence for significant benefits of such efforts is somewhat limited at present, sufficient utilisation at broad spatial and temporal scales is much more likely to be transformative within urban regions. Greater governmental support at all levels should be provided to encourage such initiatives, where it is not already present; despite this becoming more common in planning and design policy, it is generally not well ingrained, incentivised or regulated (Hostetler *et al.*, 2011). This will require greater transdisciplinary research and implementation – never an easy task, but an essential one for real progress in urban socio-ecological systems (e.g. Pickett *et al.*, 2011).

Urban ecologists remain at the forefront in several aspects of ecological innovation. The development and implementation of novel forms of ecological engineering, to find ways of melding some form of nature with our human environment, is one area. The second is in moving away from the traditional model of nature conservation, which is the preservation of ecosystems in relatively non-dynamic, familiar states, to a recognition that our new (ecologically speaking) and dynamic urban systems will contain novel assemblages. This does not mean that there is no value in ensuring preservation of remnant ecosystems and species in the urban environment, but rather that this is far from the whole story. Likewise, although efforts to restrict and control non-native species are crucial, particularly with regard to invasive species, there is little point in trying to maintain or re-create entirely native communities, and restoration efforts that hold too close to this line are likely to be unsuccessful. The key intention should be to ensure that essential ecosystem services can be maintained in urban regions, so that the urban environment functions as well as it can, and helps to relieve pressure on non-urban ecosystems. Progress perhaps lies in guiding, managing and engineering novel ecosystems with this aim in mind. Charles Elton's (1958, p. 145) oft-quoted line on the future of conservation and the necessity

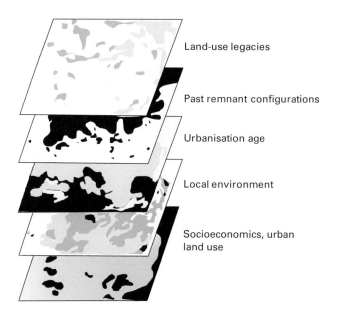

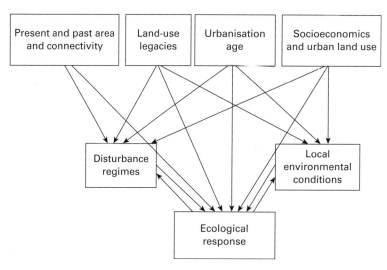

Figure 9.2 The Dynamic Urban Framework, highlighting how both spatial and temporal layers of land-use data should be considered in relation to current ecological patterns and processes, and the varying spatio-temporal factors that may determine ecological response in urban ecosystems. (Modified from Ramalho and Hobbs, 2012; reproduced with permission.)

of compromise ('reconciliation', in this case) surely applies most aptly to urban ecosystems:

> Unless one merely thinks man was intended to be an all-conquering and sterilizing power in the world, there must be…some wise principle of co-existence between

man and nature, even if it has to be a modified kind of man and a modified kind of nature. This is what I understand by conservation.

9.6 Urban research

All of our understanding of urban ecosystems and urban ecology, and our responses to urbanisation, is based on the ongoing body of urban ecology research that has taken place in recent decades (Figure 1.1). There are a myriad of important areas for further research in urban ecosystems and ecology, too exhaustive to discuss here (though see Redman *et al.*, 2004; James *et al.*, 2009; Wenger *et al.*, 2009; Jim, 2011). What we consider the most significant themes are as follows:

1   *Urban socio-ecological systems.* Urban regions are perhaps the most definitive socio-ecological systems (Figure 1.3), and there has been increasing work recognising the inter-relationship between social and ecological components of urban ecosystems (e.g. Pickett *et al.*, 2011). Much of the uncertainty in urban ecology comes from the complexity of both social and ecological systems, and how they interact over different space and timescales (e.g. Ramalho and Hobbs, 2012; McDonnell *et al.*, 2012). Greater exploration of such interrelations within urban ecosystems will remain a major aspect of urban ecology, and will require extensive interdisciplinary and transdisciplinary work. The LTER sites set up in the USA (Box 1.1) are fundamental to this process, but wider scrutiny is required.

2   *Landscape ecology and holistic research.* Although the principles of landscape ecology are increasingly being applied to urban regions, much research into both fundamental landscape processes and those specific to urban systems (e.g. the characteristics and functioning of the built matrix in shaping ecological patterns and processes) remains to be done. This is necessary to ensure that urban conservation and planning efforts are performed at appropriate scales, with the right objectives and metrics used to determine success, and are therefore likely to be effective. This is another area where the Dynamic Urban Framework (Figure 9.2) may be usefully applied to help shape research investigations.

3   *Quantification of impacts of urbanisation.* Most measurements of urbanisation impacts have been based on non-urban/exurban/urban gradient analysis, or comparisons of urban areas with 'less-urban' sites that are assumed to represent preurban states (Ramalho and Hobbs, 2012). This is because urban systems have developed so rapidly, often on already modified land, and because urban ecology has been relatively late in joining the ecological fold. This is a somewhat limited research design, as it can be difficult to ensure that 'non-urban' equals 'pre-urban' in many cases. Urbanisation is an ongoing process that shows little sign of abating in many regions. Where this cannot be avoided, it does at least present opportunities for rigorous experiments to more fully allow comparisons between urbanised and pre-urban areas, rather than comparing non-urban areas and assuming that they are representative.

4   *Non-native and invasive species.* Despite invasive species being recognised as a major threat to biodiversity and ecosystem services globally, there is relatively little work on non-natives and invasive species in urban environments, perhaps because as the most less-than-natural systems in existence, they are not seen as a

cause for concern. However, the abundance of non-native and invasive species in urban regions means that they may be key sites for examining rates of introduction and establishment, detecting possibly harmful species early on, and establishing how they may respond to different environmental conditions and control. Monitoring sites established at major urban regions within each country, perhaps maintained by trained citizen scientists (see Section 7.6), would help to determine patterns of invasion in the wider landscape. Urban regions are also excellent sites for examining societal responses to non-native and invasive species, from the complexities of invasive control to perceptions and values associated with such species.

5   *Ecological engineering and reconciliation ecology.* Operating within practical limitations is essential for ecosystem management, particularly where biodiversity is concerned. Reconciliation ecology as a means for improving biodiversity in anthropogenic landscapes has been championed in recent years, given the relative lack of ecosystem preservation and restoration options in urban regions (e.g. Lundholm and Richardson, 2010; Francis and Lorimer, 2011). This remains a nascent concept, however, and much work remains to be done in determining which forms of engineering are both possible and appropriate, whether species populations will be able to persist as a result, and the overall feasibility of such efforts.

6   *Ecosystem services.* Ecosystem services as a framework for valuing ecosystems is making substantial progress in both science and application. Regardless of the ethics of placing monetary values on the services ecosystems may provide, it is at least an effective tool for highlighting the cost of artificially replacing them. Urban areas are particularly in need of services due to their relative paucity of natural ecosystems and the many demands placed on what does remain by high population densities. Ecosystem services are increasingly coming to the attention of urban planners and being recognised as an essential aspect of sustainability. Greater evaluation of how services may be provided effectively, and placing this within appropriate economic and ethical contexts for urban regions, is important for ensuring that the science is sound and that this remains on the urban planning agenda.

7   *Species responses.* Despite some detailed work on certain species or groups of organisms, there is still limited knowledge of how organisms are adapting to the challenges and opportunities presented by urban environments. Greater knowledge of this will be important not just for establishing likely trends in biodiversity and conservation strategies, but also whether natural selection is occurring, and the implications this may have for the evolution of truly urban species. This will rely heavily on genetics analyses; techniques commonly applied to conservation in other ecosystems, but less common in an urban context.

8   *A focus on urbanisation in developing countries.* The large majority of urban ecology research has focused on developed urban regions, yet perhaps the most crucial applications are in developing countries still experiencing rapid urbanisation. Greater focus on such areas is certain over the next few decades, but should be encouraged. The primary exception is China, which maintains a substantial scientific interest in urban ecosystems.

9   *Development of LTER sites*. Developing an understanding of issues relating to these research areas can be problematic, as urban ecosystems are complex and dynamic. Despite some commonalities (Shochat *et al.*, 2006) they are variable in some ecological characteristics, such as both pre- and post-urban diversity. The LTER sites in the USA (Box 1.1) were set up to help rectify this barrier to research, but there remain just two sites in the USA, which might vary significantly from other regions. The establishment of more LTER sites in different parts of the world would be a major step forward – particularly in rapidly urbanising areas such as China, where such possibilities do exist. Ideally a network of sites would be created for the most effective comparisons.

9.7 Concluding notes

Chapters 1–8 have introduced some key aspects of urban ecosystems, highlighting some of the complexities that exist for such systems and the way in which we manage them. From uncertainties over how 'urban' systems may be best defined and characterised, to questions over the ecological impacts of urbanisation and how this may best be mitigated, there is still much to be investigated. Certainly consideration of ecosystem patterns, processes and management at the landscape level is likely to be fundamental to developing a more complete understanding of urban ecosystems, as is the development of innovative new ways of encouraging the long-term co-existence of humans and other species in urban environments. Urban planning, design and conservation will increasingly incorporate these aspects of ecology in the future, as society strives towards more sustainable urban environments. Although urbanisation will continue to create environmental impacts, we can be optimistic that urban ecosystems are at last receiving the attention they deserve. They represent our contemporary human environment, and their ecological significance should not be overlooked.

Urban ecosystems are frustratingly complex and endlessly fascinating. A short textbook like this cannot cover all aspects of urban ecosystems and their ecology, but hopefully there is sufficient breadth of coverage to encourage more detailed investigation by those interested in our human environment and how it may be improved. As noted in Chapter 1, urban environments are both a major achievement of our civilisation and one of its enduring legacies. The initial ecological investigations of urban systems touched on here hopefully represent a starting point for the development of more sustainable urban ecosystems.

9.8 Chapter summary

1   Urban growth and associated impacts are likely to continue into the middle of the twenty-first century, with the majority of growth centred in Asia, Africa and South America.
2   Differential increases and decreases in urban diversity of species at local scales will continue, while global biodiversity will continue to be impacted by urbanisation.
3   Urban sustainability will be a priority for urban regions – there are many ways in which this may be enhanced, though expectations must be kept realistic.
4   The principles and practice of landscape ecology will continue to contribute to urban planning and design, though whether this will happen fast enough to

inform urban conservation and sustainability is uncertain. Incorporation of a temporal element in urban planning (so far somewhat lacking) will be crucial, and the Dynamic Urban Framework (Figure 8.2) may be a good model to facilitate this.

5   There are many important areas for urban ecological research, with some of the key ones including the links between social and ecological components; patterns and processes at the landscape scale; quantification of differential impacts of urbanisation; the roles and impacts of non-native and invasive species; ecological engineering and reconciliation ecology potential; the role and potential of an ecosystem services framework; the ways in which species may respond to urban environments and the implications this may have for natural selection; differences in urbanisation between developed and developing regions; and the creation of LTER sites and networks around the world.

9.9 Discussion questions

1   What urbanising regions are most likely to impact on global biodiversity? For example, which urban regions are located in biodiversity hotspots?
2   Why is urban sustainability emerging only slowly, given that it is a major environmental and societal issue?
3   What research areas would you prioritise for urban ecosystems, and why?

References

Adams, L.W. (2005) 'Urban wildlife ecology and conservation: a brief history of the discipline', *Urban Ecosystems*, 8: 139–156.

Ahrne, K., Bengtsson, J. and Elmqvist, T. (2009) 'Bumble bees (*Bombus* spp.) along a gradient of increasing urbanization', *PLOS ONE*, 4: e5574.

Akbari, H., Rose, L.S and Taha, H. (2003) 'Analyzing the land cover of an urban environment using high-resolution orthophotos', *Landscape and Urban Planning*, 63: 1–14.

Albrecht, H., Eder, E., Langbehn, T. and Tschiersch, C. (2011) 'The soil seed bank and its relationship to the established vegetation in urban wastelands', *Landscape and Urban Planning*, 100: 87–97.

Alexander, D. (1989) 'Urban landslides', *Progress in Physical Geography*, 13: 157–191.

Alvey, A.A. (2006) 'Promoting and preserving biodiversity in the urban forest', *Urban Forestry and Urban Greening*, 5: 195–201.

Amati, M. (2008) 'Green belts: a twentieth-century planning experiment', in M. Amati (ed.) *Urban Green Belts in the Twenty-first Century*. London: Ashgate.

Amati, M. and Yokohari, M. (2006) 'Temporal changes and local variations in the functions of London's green belt', *Landscape and Urban Planning*, 75: 125–142.

Anas, A., Arnott, R. and Small, K.A. (1998) 'Urban spatial structure', *Journal of Economic Literature*, 36: 1426–1464.

Andersson, E., Ahrne, K., Pyykonen, M. and Elmqvist, T. (2009) 'Patterns and scale relations among urbanization measures in Stockholm, Sweden', *Landscape Ecology*, 24: 1331–1339.

Angel, S., Parent, J. and Civco, D.L. (2012) 'The fragmentation of urban landscapes: global evidence of a key attribute of the spatial structure of cities, 1990–2000', *Environment and Urbanization*, 24: 249–283.

Angold, P.G., Sadler, J.P., Hill, M.O., Pullin, A., Rushton, S., Austin, K., Small, E., Wood, B., Wadsworth, R., Sanderson, R. and Thompson, K. (2006) 'Biodiversity in urban habitat patches', *Science of the Total Environment*, 360: 196–204.

Araújo, M.B. (2003) 'The coincidence of people and biodiversity in Europe', *Global Ecology and Biogeography*, 12: 5–12.

Arnold, C.L. and Gibbons, C.J. (1996) 'Impervious surface coverage: the emergence of a key environmental indicator', *Journal of the American Planning Association*, 62: 243–258.

Attrill, M., Bilton, D.T., Rowden, A.A., Rundle, S.D. and Thomas, R.M. (1999) 'The impact of encroachment and bankside development on the habitat complexity and supralittoral invertebrate communities of the Thames Estuary foreshore', *Aquatic Conservation: Marine and Freshwater Ecosystems*, 9: 237–247.

Badyaev, A.V., Young, R.L., Oh, K.P. and Addison, C. (2008) 'Evolution on a local scale: Developmental, functional, and genetic bases of divergence in bill form and associated changes in song structure between adjacent habitats', *Evolution*, 62: 1951–1964.

Baek, Y.W. and An, Y.J. (2010) 'Assessment of toxic heavy metals in urban lake sediments as related to urban stressor and bioavailability', *Environmental Monitoring and Assessment*, 171: 529–537.

Baeten, G. (2001) 'The Europeanization of Brussels and the urbanization of "Europe" – hybridizing the city: empowerment and disempowerment in the EU district', *European Urban and Regional Studies*, 8: 117–130.

Baguette, M. and Van Dyck, H. (2007) 'Landscape connectivity and animal behavior: functional grain as a key determinant for dispersal', *Landscape Ecology*, 22: 1117–1129.

Baker, P.J. and Harris, S. (2007) 'Urban mammals: what does the future hold? An analysis of the factors affecting patterns of use of residential gardens in Great Britain', *Mammal Review*, 37: 297–315.

Baker, P.J., Funk, S.M., Harris, S. and White, P.C.L. (2000) 'Flexible spatial organization of urban foxes, *Vulpes vulpes*, before and during an outbreak of sarcoptic mange', *Animal Behaviour*, 59: 127–146.

Baldasano, J.M., Valera, E. and Jimenez, P. (2003) 'Air quality data from large cities', *Science of the Total Environment*, 307: 141–165.

Bark, R.H., Osgood, D.E., Colby, B.G., Katz, G. and Stromberg, J. (2009) 'Habitat preservation and restoration: do homebuyers have preferences for quality habitat?', *Ecological Economics*, 68: 1465–1475.

Barkham, P. (2009) 'Is it time to start culling parakeets?', *The Guardian*, Monday 12 October. Online. Available: http://www.guardian.co.uk/environment/2009/oct/12/ring-necked-parakeet-cull (accessed 29 August 2012).

Barthel, S., Folke, C. and Colding, J. (2010) 'Social-ecological memory in urban gardens: retaining the capacity for management of ecosystem services', *Global Environmental Change – Human and Policy Dimensions*, 20: 255–265.

Bates, A.J., Mackay, R., Greswell, R.B. and Sadler, J.P. (2009) 'SWITCH in Birmingham, UK: experimental investigation of the ecological and hydrological performance of extensive green roofs', *Reviews in Environmental Science and Bio-Technology*, 8: 295–300.

Bax, N., Williamson, A., Aguero, M., Gonzalez, E. and Geeves, W. (2003) 'Marine invasive alien species: a threat to global biodiversity', *Marine Policy*, 27: 313–323.

Beben, D. (2012) 'Crossings for animals: an effective method of wild fauna conservation', *Journal of Environmental Engineering and Landscape Management*, 20: 86–96.

Bengston, D.N. and Youn, Y.C. (2006) 'Urban containment policies and the protection of natural areas: the case of Seoul's greenbelt', *Ecology and Society*, 11: 3.

Bentley, M.G. (2012) '*Eriocheir sinensis* H. Milne-Edwards (Chinese mitten crab)', in R.A. Francis (ed.) *A Handbook of Global Freshwater Invasive Species*. Abingdon, Oxon: Earthscan.

Bento, A.M., Cropper, M.L., Mobarak, A.M. and Vinha, K. (2005) 'The effects of urban spatial structure on travel demand in the United States', *Review of Economics and Statistics*, 87: 466–478.

Berry, B.J.L. (1990) 'Urbanization', in B.L. Turner III, W.C. Clark, R.W. Kates, J.F. Richards, J.T. Mathews and W.B. Meyer (eds) *The Earth as Transformed by Human Action*. Cambridge: Cambridge University Press.

Bigirimana, J., Bogaert, J., De Canniere, C., Lejoly, J. and Parmentier, I. (2011) 'Alien plant species dominate the vegetation in a city of Sub-Saharan Africa', *Landscape and Urban Planning*, 100: 251–267.

Bignal, K.L., Ashmore, M.R., Headley, A.D., Stewart, K. and Weigert, K. (2007) 'Ecological impacts of air pollution from road transport on local vegetation', *Applied Geochemistry*, 22: 1265–1271.

Blacksmith Institute (2007) *The World's Worst Polluted Places: the Top Ten of the Dirty Thirty*, New York: Blacksmith Institute. Online. Available: http://www.blacksmithinstitute.org/wwpp2007/finalReport2007.pdf (accessed 12 August 2012).

Boase, C. (2008) 'Bed bugs (Hemiptera: Cimicidae): an evidence-based analysis of the current situation', in W.H. Robinson and D. Bajomi (eds) *Proceedings of the Sixth International Conference on Urban Pests*. Veszprém, Hungary: OOK-Press Kft.

Bokony, V., Kulcsar, A. and Liker, A. (2010) 'Does urbanization select for weak competitors in house sparrows?', *Oikos*, 119: 437–444.

Boone, C.G. (2008) 'Environmental justice as process and new avenues for research', *Environmental Justice*, 3: 149–154.

Booth, D.B. (2005) 'Challenges and prospects for restoring urban streams: a perspective from the Pacific Northwest of North America', *Journal of the North American Benthological Society*, 24: 724–737.

Bornkamm, R. (2007) 'Spontaneous development of urban woody vegetation on differing soils', *Flora*, 202: 695–704.

Bottrill, M.C., Joseph, L.N., Carwardine, J., Bode, M., Cook, C.N., Game, E.T., Grantham, H., Kark, S., Linke, S., McDonald-Madden, E., Pressey, R.L., Walker, S., Wilson, K.A. and Possingham, H.P. (2008) 'Is conservation triage just smart decision making?', *Trends in Ecology and Evolution*, 25: 649–654.

Boyko, C.T., Gaterell, M.R., Barber, A.R.G., Brown, J., Bryson, J.R., Butler, D., Caputo, S., Caserio, M., Coles, R., Cooper, R., Davies, G., Farmani, R., Hale, J., Hales, A.C., Hewitt, C.N., Hunt, D.V.L., Jankovic, L., Jefferson, I.,Leach, J.M., Lombardi, D.R., MacKenzie, A.R., Memon, F.A., Pugh, T.A.M., Sadler, J.P., Weingaertner, C., Whyatt, J.D. and Rogers, C.D.F. (2012) 'Benchmarking sustainability in cities: the role of indicators and future scenarios', *Global Environmental Change – Human and Policy Dimensions*, 22: 245–254.

Bradman, A., Chevrier, J., Tager, I., Lipsett, M., Sedgwick, J., Macher, J., Vargas, A.B., Cabrera, E.B., Camacho, J.M., Weldon, R., Kogut, K., Jewell, N.P. and Eskenazi, B. (2005) 'Association of housing disrepair indicators with cockroach and rodent infestations in a cohort of pregnant Latina women and their children', *Environmental Health Perspectives*, 113: 1795–1801.

Bradshaw, A.D. (1997) 'What do we mean by restoration?', in K.M. Urbanska, N.R. Webb and P.J. Edwards (eds) *Restoration Ecology and Sustainable Development*. Cambridge: Cambridge University Press.

Bramley, G., Dempsey, N., Power, S., Brown, C. and Watkins, D. (2009) 'Social sustainability and urban form: evidence from five British cities', *Environment and Planning A*, 41: 2125–2142.

Brandes, D. (2005) 'Diversity of cormophytes on inner city railway lands', *Tuexenia*, 25: 269–284.

Braun, B. (2005) 'Environmental issues: writing a more-than-human urban geography', *Progress in Human Geography*, 29: 635–650.

Breuste, J., Niemelä, J. and Snep, R.P.H. (2008) 'Applying landscape ecological principles in urban environments', *Landscape Ecology*, 23: 1139–1142.

Browne, M.A. and Chapman, M.G. (2011) 'Ecologically informed engineering reduces loss of intertidal biodiversity on artificial shorelines', *Environmental Science and Technology*, 45: 8204–8207.

Bryant, M.M. (2006) 'Urban landscape conservation and the role of ecological greenways at local and metropolitan scales', *Landscape and Urban Planning*, 76: 23–44.

Bühler, O., Balder, H. and Kristoffersen, P. (2009) 'Establishment of urban trees', *CAB Reviews: Perspectives in Agriculture, Veterinary Science, Nutrition and Natural Resources*, 4: 059.

Buijs, A.E., Elands, B.H.M. and Langers, F. (2009) 'No wilderness for immigrants: cultural differences in images of nature and landscape preferences', *Landscape and Urban Planning*, 91: 113–123.

Byrne, J.A. (2012) 'When green is white: the cultural politics of race, nature and social exclusion in a Los Angeles urban national park', *Geoforum*, 43: 595–611.

Byrne, K. and Nichols, R.A. (1999) '*Culex pipiens* in London Underground tunnels: differentiation between surface and subterranean populations', *Heredity*, 82: 7–15.

Cadenasso, M.L., Pickett, S.T.A. and Schwarz, K. (2007) 'Spatial heterogeneity in urban ecosystems: reconceptualizing land cover and a framework for classification', *Frontiers in Ecology and Environment*, 5: 80–88.

Cao, S., Xu, C., Ye, H., Zhan, Y. and Gong, C. (2010) 'The use of air bricks for planting roadside vegetation: a new technique to improve landscaping of steep roadsides in China's Hubei Province', *Ecological Engineering*, 36: 697–702.

Carbó-Ramírez, P. and Zuria, I. (2011) 'The value of small urban greenspaces for birds in a Mexican city', *Landscape and Urban Planning*, 100: 213–222.

Cariñanos, P. and Casares-Porcel, M. (2011) 'Urban green zones and related pollen allergy: a review. Some guidelines for designing spaces with low allergy impact', *Landscape and Urban Planning*, 101: 205–214.

Carpaneto, G.M., Mazziotta, A., Coletti, G., Luiselli, L. and Audisio, P. (2010) 'Conflict between insect conservation and public safety: the case study of a saproxylic beetle (*Osmoderma eremita*) in urban parks', *Journal of Insect Conservation*, 14: 555–565.

Carpenter, S.R. and Cottingham, K.L. (1997) 'Resilience and restoration of lakes', *Conservation Ecology*, 1: 2.

Carpenter, S.R., Mooney, H.A., Agard, J., Capistrano, D., DeFries, R.S., Díaz, S., Dietz, T., Duraiappah, A.K., Oteng-Yeboah, A., Pereira, H.M., Perrings, C., Reid, W.V., Sarukhanm, J., Scholes, R.J. and Whyte, A. (2009) 'Science for managing ecosystem services: beyond the Millennium Ecosystem Assessment', *Proceedings of the National Academy of Sciences of the USA*, 106: 1305–1312.

Carroll, D.G. and Jackson, C.R. (2009) 'Observed relationships between urbanization and riparian cover, shredder abundance, and stream leaf litter standing crops', *Fundamental and Applied Limnology*, 173: 213–225.

Carter, T. and Fowler, L. (2008) 'Establishing green roof infrastructure through environmental policy instruments', *Environmental Management*, 42: 151–164.

Catalan, B., Sauri, D. and Serra, P. (2008) 'Urban sprawl in the Mediterranean? Patterns of growth and change in the Barcelona Metropolitan Region 1993–2000', *Landscape and Urban Planning*, 85: 174–184.

Cetin, M. (2009) 'A satellite based assessment of the impact of urban expansion around a lagoon', *International Journal of Environmental Science and Technology*, 6: 579–590.

Chace, J.F. and Walsh, J.J. (2006) 'Urban effects on native avifauna: a review', *Landscape and Urban Planning*, 74: 46–69.

Chadwick, M.A., Huryn, A.D., Benke, A.C. and Dobberfuh, D.R. (2010) 'Coarse organic matter dynamics in urban tributaries of the St. Johns River, Florida', *Freshwater Forum*, 28: 77–93.

Chadwick, M.A., Thiele, J.E., Huryn, A.D., Benke, A.C. and Dobberfuhl, D.R. (2012) 'Effects of urbanization on macroinvertebrates in tributaries of the St. Johns River, Florida, USA', *Urban Ecosystems*, 15: 347–365.

Chamberlain, D.E., Toms, M.P., Cleary-McHarg, R. and Banks, A.N. (2007) 'House sparrow (*Passer domesticus*) habitat use in urbanized landscapes', *Journal of Ornithology*, 148: 453–462.

Chapman, M.G. and Underwood, A.J. (2011) 'Evaluation of ecological engineering of "armoured" shorelines to improve their value as habitat', *Journal of Experimental Marine Biology and Ecology*, 400: 302–313.

Chipchase, A. (1999) 'Ecological characteristics of the flora of urban derelict sites in London', *London Naturalist*, 78: 19–34.

Clark, G.F. and Johnston, E.L. (2009) 'Propagule pressure and disturbance interact to overcome biotic resistance of marine invertebrate communities', *Oikos*, 118: 1679–1686.

Clarke, K.M., Fisher, B.L. and LeBuhn, G. (2008) 'The influence of urban park characteristics on ant (Hymenoptera, Formicidae) communities', *Urban Ecosystems*, 11: 317–334.

Clarkson, B.D., Wehi, P.M. and Brabyn, L.K. (2007) 'A spatial analysis of indigenous cover patterns and implications for ecological restoration in urban centres, New Zealand', *Urban Ecosystems*, 10: 441–457.

Clayton, S. and Myers, G. (2009) *Conservation Psychology: Understanding and Promoting Human Care for Nature*. Chichester: Wiley-Blackwell.

Clergeau, P., Tapko, N. and Fontaine, B. (2011) 'A simplified method for conducting ecological studies of land snail communities in urban landscapes', *Ecological Research*, 26: 515–521.

Codd, G.A. (1995) 'Cyanobacterial toxins: occurrence, properties and biological significance', *Water Science and Technology*, 32: 149–156.

Cohen, A.J., Anderson, H.R., Ostro, B., Pandey, K.D., Krzyzanowski, M., Kuenzli, N., Gutschmidt, K., Pope, C.A., Romieu, I., Samet, J.M. and Smith, K.R. (2004) 'Mortality impacts of urban air pollution', in M. Ezzati, A.D. Lopez, A. Rodgers and C.U.J.L. Murray (eds) *Comparative Quantification of Health Risks: Global and Regional Burden of Disease due to Selected Major Risk Factors*. Geneva: World Health Organization.

Colding, J. (2007) 'Ecological land-use complementation for building resilience in urban ecosystems', *Landscape and Urban Planning*, 81: 46–55.

Colding, J. and Folke, C. (2009) 'The role of golf courses in biodiversity conservation and ecosystem management', *Ecosystems*, 12: 191–206.

Colding, J., Lundberg, J., Lundberg, S. and Andersson, E. (2009) 'Golf courses and wetland fauna', *Ecological Applications*, 19: 1481–1491.

Collier, C.G. (2006) 'The impact of urban areas on weather', *Quarterly Journal of the Royal Meteorological Society*, 132: 1–25.

Connell, J.H. (1978) 'Diversity in tropical rain forests and coral reefs – high diversity or trees and corals is maintained only in a non-equilibrium state', *Science*, 199: 1302–1310.

Connery, K. (2009) 'Biodiversity and urban design: seeking an integrated solution', *Journal of Green Building*, 4: 23–38.

Contesse, P., Hegglin, D., Gloor, S., Bontadina, F. and Deplazes, P. (2004) 'The diet of urban foxes (*Vulpes vulpes*) and the availability of anthropogenic food in the city of Zurich, Switzerland', *Mammal Biology*, 69: 81–95.

Cook, E.M., Hall, S.J. and Larson, K.L. (2012) 'Residential landscapes as social-ecological systems: a synthesis of multi-scalar interactions between people and their home environment', *Urban Ecosystems*, 15: 19–52.

Cooper, C.B., Dickinson, J., Phillips, T. and Bonney, R. (2007) 'Citizen science as a tool for conservation in residential ecosystems', *Ecology and Society*, 12: 11.

Cooper, C.B., Loyd, K.A.T., Murante, T., Savoca, M. and Dickinson, J. (2012) 'Natural history traits associated with detecting mortality within residential bird communities: can citizen science provide insights?', *Environmental Management*, 50: 11–20.

Cornelis, J. and Hermy, M. (2004) 'Biodiversity relationships in urban and suburban parks in Flanders', *Landscape and Urban Planning*, 69: 385–401.

Cornell, K.L., Kight, C.R., Burdge, R.B., Gunderson, A.R., Hubbard, J.K., Jackson, A.K., LeClerc, J.E., Pitts, M.L., Swaddle, J.P. and Cristol, D.A. (2011) 'Reproductive success of Eastern Bluebirds (*Sialis sialis*) on suburban golf courses', *Auk*, 128: 577–586.

Coutts, A.M., Tapper, N.J., Beringer, J., Loughnan, M. and Demuzere, M. (in press) 'Watering our cities: the capacity for water sensitive urban design to support urban cooling and improve human thermal comfort in the Australian context', *Progress in Physical Geography*.

Craul, P.J. (1992) *Urban Soil in Landscape Design*. New York: John Wiley & Sons.

Croci, S., Butet, A., Georges, A., Aguejdad, R. and Clergeau, P. (2008) 'Small urban woodlands as biodiversity conservation hot-spot: a multi-taxon approach', *Landscape Ecology*, 23: 1171–1186.

Crooks, K.R., Suarez, A.V. and Bolger, D.T. (2004) 'Avian assemblages along a gradient of urbanization in a highly fragmented landscape', *Biological Conservation*, 115: 451–462.

Csuzdi, C., Pavlíček, T. and Nevo, E. (2008) 'Is *Dichogaster bolaui* (Michaelsen, 1891) the first domicile earthworm species?', *European Journal of Soil Biology*, 44: 198–201.

Currie, W.S. (2011) 'Units of nature or processes across scales? The ecosystem concept at age 75', *New Phytologist*, 190: 21–34.

Daily, G.C., Polasky, S., Goldstein, J., Kareiva, P.M., Mooney, H.A., Pejchar, L., Ricketts, T.H., Salzman, J. and Shallenberger, R. (2009) 'Ecosystem services in decision making: time to deliver', *Frontiers in Ecology and Environment*, 7: 21–28.

Dallimer, M., Irvine, K.N., Skinner, A.M.J., Davies, Z.G., Rouquette, J.R., Maltby, L.L., Warren, P.H., Armsworth, P.R. and Gaston, K.J. (2012) 'Biodiversity and the feel-good factor: understanding associations between self-reported human well-being and species richness', *BioScience*, 62: 47–55.

Darlington, A. (1981) *Ecology of Walls*. London: Heinemann Educational Books.

Davies, T.W., Bennie, J. and Gaston, K.J. (2012) 'Street lighting changes the composition of invertebrate communities', *Biology Letters*, 8: 764–767.

Davies, Z.G., Fuller, R.A., Loram, A., Irvine, K.N., Sims, V. and Gaston, K.J. (2009) 'A national scale inventory of resource provision for biodiversity within domestic gardens', *Biological Conservation*, 142: 761–771.

Davis, M., Chew, M.K., Hobbs, R.J., Lugo, A.E., Ewel, J.J., Vermeij, G.J., Brown, J.H., Rosenzweig, M.L., Gardener, M.R., Carroll, S.P., Thompson, K., Pickett, S.T.A., Stromberg, J.C., Del Tredici, P., Suding, K.N., Ehrenfeld, J.G., Grime, J.P., Mascaro, J. and Briggs, J.C. (2011) 'Don't judge species on their origins', *Nature*, 474: 153–154.

Davison, A. and Ridder, B. (2006) 'Turbulent times for urban nature: conserving and re-inventing nature in Australian cities', *Australian Zoologist*, 33: 306–314.

Davison, J., Huck, M., Delahay, R.J. and Roper, T.J. (2009) 'Restricted ranging behaviour in a high-density population of urban badgers', *Journal of Zoology*, 277: 45–53.

Dawe, G.F.M. (2010) 'Street trees and the urban environment', in I. Douglas, D. Goode, M. Houck and R. Wang (eds) *The Routledge Handbook of Urban Ecology*, London: Routledge.

de Neef, D., Stewart, G.H. and Meurk, C.D. (2008) 'Urban biotopes of Aotearoa New Zealand (URBANZ) (III): spontaneous urban wall vegetation in Christchurch and Dunedin', *Phyton*, 48: 133–154.

Diaz, R.J. and Rosenberg, R. (2008) 'Spreading dead zones and consequences for marine ecosystems', *Science*, 321: 926–929.

Dietzel, C., Herold, M., Hemphill, J.J. and Clarke, K.C. (2005) 'Spatio-temporal dynamics in California's central valley: empirical links to urban theory', *International Journal of Geographical Information Science*, 19: 175–195.

Dorotovicova, C. and Ot'ahel'ova, H. (2008) 'The influence of anthropogenic factors on the structure of aquatic macrophytes vegetation in the Hurbanovsky canal (South Slovakia)', *Archiv für Hydrobiologie*, 166: 81–90.

Drinnan, I.N. (2005) 'The search for fragmentation thresholds in a southern Sydney suburb', *Biological Conservation*, 124: 339–349.

Duerksen, C.J., Elliott, D.L., Hobbs, N.T., Johnson, E. and Miller, J.R. (1997) *Habitat Protection Planning: Where the Wild Things Are*. American Planning Association, Planning Advisory Service, Report number 470/471.

Dunn, R.R. (2010) 'Global mapping of ecosystem disservices: the unspoken reality that nature sometimes kills us', *Biotropica*, 42: 555–557.

Dunnett, N. and Kingsbury, N. (2008) *Planting Green Roofs and Living Walls*. London: London: Timber Press.

Dunnett, N., Nagase, A. and Hallam, A. (2008) 'The dynamics of planted and colonising species on a green roof over six growing seasons 2001–2006: influence of substrate depth', *Urban Ecosystems*, 11: 373–384.

Ebeling, S.K., Welk, E., Auge, H. and Bruelheide, H. (2008) 'Predicting the spread of an invasive plant: combining experiments and ecological niche model', *Ecography*, 31: 709–719.

Edmondson, J., Davies, Z.G., McCormack, S.A., Gaston, K.J. and Leake, J.R. (2011) 'Are soils in urban ecosystems compacted? A citywide analysis', *Biology Letters*, 7: 771–774.

Elmore, A.J. and Kaushal, S.S. (2008) 'Disappearing headwaters: patterns of stream burial due to urbanization', *Frontiers in Ecology and the Environment*, 6: 308–312.

Elton, C.S. (1958) *The Ecology of Invasions by Animals and Plants*. Chicago, IL: University of Chicago Press.

Emilsson, T. (2008) 'Vegetation development on extensive vegetated green roofs: influence of substrate composition, establishment method and species mix', *Ecological Engineering*, 33: 265–277.

Erickson, D.L. (2004) 'The relationship of historic city form and contemporary greenway implementation: a comparison of Milwaukee, Wisconsin (USA) and Ottawa, Ontario (Canada)', *Landscape and Urban Planning*, 68: 199–221.

Escobedo, F.J., Kroeger, T. and Wagner, J.E. (2011) 'Urban forests and pollution mitigation: analyzing ecosystem services and disservices', *Environmental Pollution*, 159: 2078–2087.

Evans, J. (2004) 'What is local about local environmental governance? Observations from the local biodiversity action planning process', *Area*, 36: 270–279.

Evans, K.L., Gaston, K.J., Sharp, S.P., McGowan, A. and Hatchwell, B.J. (2009) 'The effect of urbanisation on avian morphology and latitudinal gradients in body size', *Oikos* 118: 251–259.

Evans, K.L., Chamberlain, D.E., Hatchwell, B.J., Gregory, R.D. and Gaston, K.J. (2011) 'What makes an urban bird?', *Global Change Biology*, 17: 32–44.

Everaars, J., Strohbach, M.W., Gruber, B. and Dormann, C.F. (2011) 'Microsite conditions dominate habitat selection of the red mason bee (*Osmia bicornis*, Hymenoptera: Megachilidae) in an urban environment: A case study from Leipzig, Germany', *Landscape and Urban Planning*, 103: 15–23.

Everard, M. and Moggridge, H.L. (2012) 'Rediscovering the value of urban rivers', *Urban Ecosystems*, 15: 293–314.

Eyre, M.D., Luff, M.L. and Woodward, J.C. (2003) 'Beetles (Coleoptera) on brownfield sites in England: an important conservation resource?', *Journal of Insect Conservation*, 7: 223–231.

Faghih, N. and Sadeghy, A. (2012) 'Persian gardens and landscapes', *Architectural Design*, 82: 38–51.

Faiers, A. and Bailey, A. (2005) 'Evaluating canalside hedgerows to determine future interventions', *Journal of Environmental Management*, 74: 71–78.

Feng, H., Cochran, J.K., Lwiza, H., Brownawell, B.J. and Hirschberg, D.J. (1998) 'Distribution of heavy metal and PCB contaminants in the sediments of an urban estuary: the Hudson River', *Marine Environmental Research*, 45: 69–88.

Ferguson, B.K. (2005) *Porous Pavements*. Boca Raton, FL: CRC Press.

Fernandez-Juricic, E. (2004) 'Spatial and temporal analysis of the distribution of forest specialists in an urban-fragmented landscape (Madrid, Spain): implications for local and regional bird conservation', *Landscape and Urban Planning*, 69: 17–32.

Fernandez-Juricic, E. and Jokimaki, J. (2001) 'A habitat island approach to conserving birds in urban landscapes: case studies from southern and northern Europe', *Biodiversity and Conservation*, 10: 2023–2043.

Ferris, J., Norman, C. and Sempik, J. (2001) 'People, land and sustainability: community gardens and the social dimension of sustainable development', *Social Policy and Administration*, 35: 559–568.

FitzGibbon, S.I., Putland, D.A. and Goldizen, A.W. (2007) 'The importance of functional connectivity in the conservation of a ground-dwelling mammal in an urban Australian landscape', *Landscape Ecology*, 22: 1513–1525.

Flavell, N. (2003) 'Urban allotment gardens in the eighteenth century: the case of Sheffield', *Agricultural History Review*, 51: 95–106.

FLL (2002) *Guidelines for the Planning, Execution and Upkeep of Green Roof Sites* (English ed.). Bonn: Forschungsgesellschaft Landschaftsentwicklung Landschaftsbau.

Foley, S.M., Price, S.J. and Dorcas, M.E. (2012) 'Nest-site selection and nest depredation of semi-aquatic turtles on golf courses', *Urban Ecosystems*, 15: 489–497.

Folke, C., Jansson, A., Larsson, J. and Costanza, R. (1997) 'Ecosystem appropriation by cities', *Ambio*, 26: 167–172.

Fontana, C.S., Burger, M.I. and Magnusson, W.E. (2011) 'Bird diversity in a subtropical South-American City: effects of noise levels, arborisation and human population density', *Urban Ecosystems*, 14: 341–360.

Forman, R.T.T. (1996) *Land Mosaics: The Ecology of Landscapes and Regions*. Cambridge: Cambridge University Press.

Forman, R.T.T. (2008) *Urban Regions: Ecology and Planning Beyond the City*. Cambridge: Cambridge University Press.

Forman, R.T.T. and Godron, M. (1986) *Landscape Ecology*. Chichester: John Wiley & Sons.

Forman, R.T.T. and Deblinger, R.D. (2000) 'The ecological road-effect zone of a Massachusetts (USA) suburban highway', *Conservation Biology*, 14: 36–46.

Forman, R.T.T., Sperling, D., Bissonette, J.A., Clevenger, A.P., Cutshall, C.D., Dale, V.H., Fahrig, L., France, R., Goldman, C.R., Heanue, K., Jones, J.A., Swanson, F.J., Turrentine, T. and Winter, T.C. (2003) *Road Ecology: Science and Solutions*. Washington, DC: Island Press.

Francis, R.A. (2009a) 'Perspectives on the potential for reconciliation ecology in urban rivers-capes', *CAB Reviews: Perspectives in Agriculture, Veterinary Science, Nutrition and Natural Resources*, 4: art 73.

Francis, R.A. (2009b) 'Ecosystem prediction and management', in N. Castree, D. Demeritt, D. Liverman and B. Rhoads (eds) *A Companion to Environmental Geography*. London: Blackwell.

Francis, R.A. (2011) 'Wall ecology: a frontier for urban biodiversity and ecological engineering', *Progress in Physical Geography*, 35: 43–63.

Francis, R.A. (2012) 'Positioning urban rivers within urban ecology', *Urban Ecosystems*, 15: 285–291.

Francis, R.A. and Hoggart, S.P.G. (2009) 'Urban river wall habitat and vegetation: observations from the River Thames through central London', *Urban Ecosystems*, 12: 468–485.

Francis, R.A. and Goodman, M.K. (2010) 'Post normal science and the art of nature conservation', *Journal for Nature Conservation*, 18: 89–105.

Francis, R.A. and Chadwick, M.A. (2011) 'Invasive alien species in freshwater ecosystems: a brief overview', in R.A. Francis (ed.) *A Handbook of Global Freshwater Invasive Species*. London: Routledge.

Francis, R.A. and Lorimer, J. (2011) 'Urban reconciliation ecology: the potential of living roofs and walls', *Journal of Environmental Management*, 92: 1429–1437.

Francis, R.A. and Chadwick, M.A. (2012) 'What makes a species synurbic?', *Applied Geography*, 32: 514–521.

Francis, R.A. and Hoggart, S.P.G. (2012) 'The flora of urban river wallscapes', *River Research and Applications*, 28: 1200–1216.

Francis, R.A. and Pyšek, P. (2012) 'Management of freshwater invasive alien species', in R.A. Francis (ed.) *A Handbook of Global Freshwater Invasive Species*. London: Routledge.

Francis, R.A., Lorimer, J. and Raco, M. (2012) 'Urban ecosystems as "natural" homes for biogeographical boundary crossings', *Transactions of the Institute of British Geographers*, 37: 183–190.

Friese, K., Schmidt, G., Carvalho, de, Lena, J., Nalini, H.A. and Zachmann, D.W. (2010) 'Anthropogenic influence on the degradation of an urban lake: the Pampulha reservoir in Belo Horizonte, Minas Gerais, Brazil', *Limnologica*, 40: 114–125.

Fuller, R.A. and Gaston, K.J. (2009) 'The scaling of green space coverage in European cities', *Biology Letters*, 5: 352–355.

Fuller, R.A., Warren, P.H., Armsworth, P.R., Barbosa, O. and Gaston, K.J. (2008) 'Garden bird feeding predicts the structure of urban avian assemblages', *Diversity and Distributions*, 14: 131–137.

Gaffin, S.R., Khanbilvardi, R. and Rosenzweig, C. (2009) 'Development of a green roof environmental monitoring and meteorological network in New York City', *Sensors*, 9: 2647–2660.

Gagne, S.A. and Fahrig, L. (2010) 'The trade-off between housing density and sprawl area: minimising impacts to forest breeding birds', *Basic and Applied Ecology*, 11: 723–733.

Gallagher, F.J., Pechmann, I., Holzapfel, C. and Grabosky, J. (2011) 'Altered vegetative assemblage trajectories within an urban brownfield', *Environmental Pollution*, 159: 1159–1166.

Galluzzi, G., Eyzaguirre, P. and Negri, V. (2010) 'Home gardens: neglected hotspots of agro-biodiversity and cultural diversity', *Biodiversity and Conservation*, 19: 3635–3654.

Gaston, K.J., Warren, P.H., Thompson, K. and Smith, R.M. (2005a) 'Urban domestic gardens (IV): the extent of the resource and its associated features', *Biodiversity and Conservation*, 14: 3327–3349.

Gaston, K.J., Smith, R.M., Thompson, K. and Warren, P.H. (2005b) 'Urban domestic gardens (II): experimental tests of methods for increasing biodiversity', *Biodiversity and Conservation*, 14: 395–413.

Gaston, K.J., Fuller, R.A., Loram, A., MacDonald, C., Power, S. and Dempsey, N. (2007) 'Urban domestic gardens (XI): variation in urban wildlife gardening in the United Kingdom', *Biodiversity and Conservation*, 16: 3227–3238.

Gates, C. (2003) *Ancient Cities: The Archaeology of Urban Life in the Ancient Near East and Egypt, Greece and Rome*. New York: Routledge.

George, S.L. and Crooks, K.R. (2006) 'Recreation and large mammal activity in an urban nature reserve', *Biological Conservation*, 133: 107–117.

Gerlach, G. and Musolf, K. (2000) 'Fragmentation of landscape as a cause for genetic subdivision in bank voles', *Conservation Biology*, 14: 1066–1074.

Gessner, M.O. and Chauvet, E. (2002) 'A case for using litter breakdown to assess functional stream integrity', *Ecological Applications*, 12: 498–510.

Gilbert, O.L. (1992) *Rooted in Stone: The Natural Flora of Urban Walls*. Peterborough: English Nature.

Giuliano, W.M. and Accamando, A.K. (2004) 'Lepidoptera–habitat relationships in urban parks', *Urban Ecosystems*, 7: 361–370.

Gleason, K.L. (1994) 'Porticus Pompeiana: a new perspective on the first public park of ancient Rome', *Journal of Garden History*, 14: 13–27.

Gledhill, D.G., James, P. and Davies, D.H. (2008) 'Pond density as a determinant of aquatic species richness in an urban landscape', *Landscape Ecology*, 23: 1219–1230.

Glista, D.J., DeVault, T.L. and DeWoody, J.A. (2009) 'A review of mitigation measures for reducing wildlife mortality on roadways', *Landscape and Urban Planning*, 91: 1–7.

Gloor, S., Bontadina, F., Hegglin, D., Deplazes, P. and Breitenmoser, U. (2001) 'The rise of urban fox populations in Switzerland', *Mammal Biology*, 66: 155–164.

Gobster, P.H. (2001) 'Visions of nature: conflict and compatibility in urban park restoration', *Landscape and Urban Planning*, 56: 35–51.

Goddard, M.A., Dougill, A.J. and Benton, T.G. (2010) 'Scaling up from gardens: biodiversity conservation in urban environments', *Trends in Ecology and Evolution*, 25: 90–98.

Godefroid, S. and Koedam, N. (2003) 'How important are large vs. small forest remnants for the conservation of the woodland flora in an urban context?', *Global Ecology and Biogeography*, 12: 287–298.

Goulder, L.H. and Kennedy, D. (1997) 'Valuing ecosystem services: philosophical bases and empirical methods', in G.C. Daily (ed.) *Nature's Services: Societal Dependence on Natural Ecosystems*. Washington, DC: Island Press.

Grant, G. (2006) 'Extensive green roofs in London', *Urban Habitats*, 4: 51–65.

Gras, L.M., Patergnani, M. and Farina, M. (2012) 'Poison-based commensal rodent control strategies in urban ecosystems: some evidence against sewer-baiting', *Ecohealth*, 9: 75–79.

Grewal, S.S. and Grewal, P.S. (2012) 'Can cities become self-reliant in food?', *Cities*, 29: 1–11.

Grime, J.P. (1977) 'Evidence for existence of 3 primary strategies in plants and its relevance to ecological and evolutionary theory', *American Naturalist*, 111: 1169–1194.

Grimm, N.B., Faeth, S.H., Golubiewski, N.E., Redman, C.L., Wu, J., Bai, X. and Briggs, J.M. (2008) 'Global change and the ecology of cities', *Science*, 319: 756–760.

Grimm, N.B., Hale, R.L., Cook, E.M. and Iwaniec, D.M. (2010) 'Urban biogeochemical flux analysis', in I. Douglas, D. Goode, M. Houck and R. Wang (eds) *The Routledge Handbook of Urban Ecology*. London: Routledge.

Grimmond, C.S.B. and Oke, T.R. (1999) 'Aerodynamic properties of urban areas derived from analysis of surface form', *Journal of Applied Meteorology*, 38: 1262–1292.

Groffman, P.M. and Pouyat, R.V. (2009) 'Methane uptake in urban forests and lawns', *Environmental Science and Technology*, 43: 5229–5235.

Groffman, P.M., Bain, D.J., Band, L.E., Belt, K.T., Brush, G.S., Grove, J.M., Pouyat, R.V., Yesilonis, I.C. and Zipperer, W.C. (2003) 'Down by the riverside: urban riparian ecology', *Frontiers in Ecology and the Environment*, 1: 315–321.

Groffman, P.M., Williams, C.O., Pouyat, R.V., Band, L.E. and Yesilonis, I.D. (2009) 'Nitrate leaching and nitrous oxide flux in urban forests and grasslands', *Journal of Environmental Quality*, 38: 1848–1860.

Guggenheim, E. (1992) 'Wall vegetation in the city of Zürich', *Berichte des Geobotanischen Institutes der Eidgenoessischen Technischen Hochschule Stiftung Ruebel Zuerich*, 58: 164–191.

Gundersen, V., Frivold, L.H., Myking, T. and Oyen, B-H. (2006) 'Management of urban recreational woodlands: the case of Norway', *Urban Forestry and Urban Greening*, 5: 73–82.

Gurnell, A., Lee, M. and Souch, C. (2007) 'Urban rivers: hydrology, geomorphology, ecology and opportunities for change', *Geography Compass*, 1: 1118–1137.

Haeckel, E. (1866) *Generelle Morphologie der Organismen*. Berlin: Verlag von Georg Reimer.

Hahs, A.K., McDonnell, M.J., McCarthy, M.A., Vesk, P.A., Corlett, R.T., Norton, B.A., Clemants, S.E., Duncan, R.P., Thompson, K., Schwartz, M.W. and Williams, N.S.G. (2009) 'A global synthesis of plant extinction rates in urban areas', *Ecology Letters*, 12: 1165–1173.

Haile, T. and Nakhla, G. (2008) 'Novel zeolite coating for protection of concrete sewers from biological sulfuric acid attack', *Geomicrobiology Journal*, 25: 322–331.

Hamby, D.M. (1996) 'Site remediation techniques supporting environmental restoration activities: a review', *Science of the Total Environment*, 191: 203–224.

Hamer, A.J., Smith, P.J. and McDonnell, M.J. (2012) 'The importance of habitat design and aquatic connectivity in amphibian use of urban stormwater retention ponds', *Urban Ecosystems*, 15: 451–471.

Hannon, E.R. and Hafernik, J.E. (2007) 'Reintroduction of the rare damselfly *Ischnura gemina* (Odonata: Coenagrionidae) into an urban California park', *Journal of Insect Conservation*, 11: 141–149.

Hansen, M.J. and Clevenger, A.P. (2005) 'The influence of disturbance and habitat on the presence of non-native plant species along transport corridors', *Biological Conservation*, 125: 249–259.

Harnik, P. (2006) 'The excellent city park system: what makes it great and how to get there', in R.H. Platt (ed.) *The Humane Metropolis: People and Nature in the 21st-Century City*. Amherst and Boston: University of Massachusetts Press.

Harper-Lore, B. and Wilson, M. (eds) (2000) *Roadside Use of Native Plants*. Washington, DC: Island Press.

Hartley, W., Uffindell, L., Plumb, A., Rawlinson, H.A., Putwain, P. and Dickinson, N.M. (2008) 'Assessing biological indicators for remediated anthropogenic urban soils', *Science of the Total Environment*, 405: 358–369.

Hartley, W., Dickinson, N.M., Riby, P. and Shutes, B. (2012) 'Sustainable ecological restoration of brownfield sites through engineering or managed natural attenuation? A case study from Northwest England', *Ecological Engineering*, 40: 70–79.

Hatt, B.E., Fletcher, T.D., Walsh, C.J. and Taylor, S.T. (2004) 'The influence of urban density and drainage infrastructure on the concentrations and loads of pollutants in small streams', *Environmental Management*, 34: 112–124.

Hawbaker, T.J., Radeloff, V.C., Hammer, R.B. and Clayton, M.K. (2005) 'Road density and landscape pattern in relation to housing density, and ownership, land cover, and soils', *Landscape Ecology*, 20: 609–625.

Hayasaka, D., Akasaka, M., Miyauchi, D., Box, E.O. and Uchida, T. (2012) 'Qualitative variation in roadside weed vegetation along an urban-rural road gradient', *Flora*, 207: 126–132.

Hedblom, M. and Söderström, B. (2008) 'Woodlands across Swedish urban gradients: status, structure and management implications', *Landscape and Urban Planning*, 84: 62–73.

Hedblom, M. and Söderström, B. (2010) 'Landscape effects on birds in urban woodlands: an analysis of 34 Swedish cities', *Journal of Biogeography*, 37: 1302–1316.

Helden, A.J. and Leather, S.R. (2004) 'Biodiversity on urban roundabouts: Hemiptera, management and the species–area relationship', *Basic and Applied Ecology*, 5: 367–377.

Hermy, M. (2010) 'Landscaped parks and open spaces', in I. Douglas, D. Goode, M. Houck and R. Wang (eds) *The Routledge Handbook of Urban Ecology*. London: Routledge.

Herold, M., Couclelis, H. and Clarke, K.C. (2005) 'The role of spatial metrics in the analysis and modeling of urban land use change', *Computers Environment and Urban Systems*, 29: 369–399.

Higgs, S., Snow, K. and Gould, E.A. (2004) 'The potential for West Nile virus to establish outside of its natural range: a consideration of potential mosquito vectors in the United Kingdom', *Transactions of the Royal Society of Tropical Medicine and Hygiene*, 98: 82–87.

Hijosa-Valsero, M., Sidrach-Cardona, R. and Becares, E. (2012) 'Comparison of interannual removal variation of various constructed wetland types', *Science of the Total Environment*, 430: 174–183.

Hinchliffe, S. and Whatmore, S. (2006) 'Living cities: towards a politics of conviviality', *Science as Culture*, 15: 123–138.

Hinchliffe, S., Kearnes, M.B., Degen, M. and Whatmore, S. (2005) 'Urban wild things: a cosmopolitical experiment', *Environment and Planning D: Society & Space*, 23: 643–658.

Hinchman, L.P. and Hinchman, S.K. (2007) 'What we owe the Romantics', *Environmental Values*, 16: 333–354.

Hobbs, R.J., Arico, S., Aronson, J., Baron, J.S., Bridgewater, P., Cramer, V.A., Epstein, P.R., Ewel, J.J., Klink, C.A., Lugo, A.E., Norton, D., Ojima, D., Richardson, D.M., Sanderson, E.W., Valladares, F., Vila, M., Zamora, R. and Zobel, M. (2006) 'Novel ecosystems: theoretical and management aspects of the new ecological world order', *Global Ecology and Biogeography*, 15: 1–7.

Hobbs, R.J., Higgs, E. and Harris, J.A. (2009) 'Novel ecosystems: implications for conservation and restoration', *Trends in Ecology and Evolution*, 24: 599–605.

Hodgkison, S., Hero, J.M. and Warnken, J. (2007) 'The efficacy of small-scale conservation efforts, as assessed on Australian golf courses', *Biological Conservation*, 135: 576–586.

Hogan, D.M. and Walbridge, M.R. (2007) 'Best management practices for nutrient and sediment retention in urban stormwater runoff', *Journal of Environmental Quality*, 36: 386–395.

Hoggart, S.P.G., Francis, R.A. and Chadwick, M.A. (2012) 'Macroinvertebrate richness on flood defence walls of the tidal River Thames', *Urban Ecosystems*, 15: 327–346.

Holland, P.G. (1972) 'The pattern of species density of old stone walls in Western Ireland', *Journal of Ecology*, 60: 799–805.

Holling, C.S. (1973) 'Resilience and stability of ecological systems', *Annual Review of Ecology and Systematics*, 4: 1–23.

Hope, D., Gries, C., Zhu, W.X., Fagan, W.F., Redman, C.L., Grimm, N.B., Nelson, A.L., Martin, C. and Kinzig, A. (2003) 'Socioeconomics drive urban plant diversity', *Proceedings of the National Academy of Sciences of the USA*, 100: 8788–8792.

Hostetler, M. and Drake, D. (2009) 'Conservation subdivisions: a wildlife perspective', *Landscape and Urban Planning*, 90: 95–101.

Hostetler, M., Allen, W. and Meurk, C. (2011) 'Conserving urban biodiversity? Creating green infrastructure is only the first step', *Landscape and Urban Planning*, 100: 369–371.

Hougner, C., Colding, J. and Soderqvist, T. (2006) 'Economic valuation of a seed dispersal service in the Stockholm National Urban Park, Sweden', *Ecological Economics*, 59: 364–374.

Howard, E. (1898) *Tomorrow: A Peaceful Path to Real Reform*. London: Sonnenschein.

Howard, J.L. and Olszewska, D. (2011) 'Pedogenesis, geochemical forms of heavy metals, and artifact weathering in an urban soil chronosequence, Detroit, Michigan', *Environmental Pollution*, 159: 754–761.

Hu, Y. and Cardoso, G.C. (2010) 'Which birds adjust the frequency of vocalizations in urban noise?', *Animal Behaviour*, 79: 863–867.

Huang, S.L. and Hsu, W.L. (2003) 'Materials flow analysis and emergysic evaluation of Taipei's urban construction', *Landscape and Urban Planning*, 63: 61–74.

Hufkens, K., Scheunders, P. and Ceulemans, R. (2009) 'Ecotones in vegetation ecology: methodologies and definitions revisited', *Ecological Research*, 24: 977–986.

Hulme, P.E. (2009) 'Trade, transport and trouble: managing invasive species pathways in an era of globalization', *Journal of Applied Ecology*, 46: 10–18.

Hunter, J.C. and Mattice, J.A. (2002) 'The spread of woody exotics into the forests of a northeastern landscape, 1938–1999', *Journal of the Torrey Botanical Society*, 129: 220–227.

Iceland, J. and Scopilliti, M. (2008) 'Immigrant residential segregation in US metropolitan areas, 1990–2000', *Demography*, 45: 79–94.

Ignatieva, M., Stewart, G.H. and Meurk, C. (2011) 'Planning and design of ecological networks in urban areas', *Landscape and Ecological Engineering*, 7: 17–25.

Imberger, S.J., Walsh, C.J. and Grace, M.R. (2008) 'More microbial activity, not abrasive flow or shredder abundance, accelerates breakdown of labile leaf litter in urban streams', *Journal of the North American Benthological Society*, 27: 549–561.

Imhoff, M.L., Bounoua, L., DeFries, R., Lawrence, W.T., Stutzer, D., Tucker, C.J. and Ricketts, T. (2004) 'The consequences of urban land transformation on net primary productivity in the United States', *Remote Sensing of Environment*, 89: 434–443.

Institution of Mechanical Engineers (2009) *Geo-engineering: Giving Us the Time to Act*. London: Institution of Mechanical Engineers.

Ishii, H.T., Manabe, T., Ito, K., Fujita, N., Imanishi, A., Hashimoto, D. and Iwasaki, A. (2010) 'Integrating ecological and cultural values toward conservation and utilization of shrine/temple forests as urban green space in Japanese cities', *Landscape and Ecological Engineering*, 6: 307–315.

Ivanov, K. and Keiper, J. (2010) 'Ant (Hymenoptera: Formicidae) diversity and community composition along sharp urban forest edges', *Biodiversity and Conservation*, 19: 3917–3933.

Jabareen, Y.R. (2006) 'Sustainable urban forms: their typologies, models, and concepts', *Journal of Planning Education and Research*, 26: 38–52.

Jackson, N.D. and Fahrig, L. (2011) 'Relative effects of road mortality and decreased connectivity on population genetic diversity', *Biological Conservation*, 144: 3143–3148.

Jacobson, C.R. (2011) 'Identification and quantification of the hydrological impacts of imperviousness in urban catchments: A review', *Journal of Environmental Management*, 92: 1438–1448.

James, P., Tzoulas, K., Adams, M.D., Barber, A., Box, J., Breuste, J., Elmqvist, T., Frith, M., Gordon, C., Greening, K.L., Handley, J., Haworth, S., Kazmierczak, A.E., Johnston, M., Korpela, K., Moretti, M., Niemelä, J., Pauleit, S., Roe, M.H., Sadler, J.P. and Thompson, C.W. (2009) 'Towards an integrated understanding of green space in the European built environment', *Urban Forestry and Urban Greening*, 8: 65–75.

Jiang, B. (2007) 'A topological pattern of urban street networks: universality and peculiarity', *Physica A – Statistical Mechanics and Its Applications*, 384: 647–655.

Jiang, M.M. and Chen, B. (2011) 'Integrated urban ecosystem evaluation and modeling based on embodied cosmic exergysic', *Ecological Modelling*, 222: 2149–2165.

Jim, C.Y. (2001) 'Managing urban trees and their soil envelopes in a contiguously developed city environment', *Environmental Management*, 28: 819–832.

Jim, C.Y. (2002) 'Planning strategies to overcome constraints on greenspace provision in urban Hong Kong', *Town Planning Review*, 73: 127–152.

Jim, C.Y. (2008) 'Urban biogeographical analysis of spontaneous tree growth on stone retaining walls', *Physical Geography*, 29: 351–373.

Jim, C.Y. (2010) 'Urban woodlands as distinctive and threatened nature-in-city patches', in I. Douglas, D. Goode, M. Houck and R. Wang (eds) *The Routledge Handbook of Urban Ecology*. London: Routledge.

Jim, C.Y. (2011) 'Holistic research agenda for sustainable management and conservation of urban woodlands', *Landscape and Urban Planning*, 100: 375–379.

Jim, C.Y. and Chen, S.S. (2003) 'Comprehensive greenspace planning based on landscape ecology principles in compact Nanjing city, China', *Landscape and Urban Planning*, 65: 95–116.

Jim, C.Y. and Chen, W.Y. (2006) 'Perception and attitude of residents toward urban green spaces in Guangzhou (China)', *Environmental Management* 38: 338–349.

Jim, C.Y. and Chen, W.Y. (2008) 'Pattern and divergence of tree communities in Taipei's main urban green spaces', *Landscape and Urban Planning*, 84: 312–323.

Jim, C.Y. and Chen, W.Y. (2009) 'Value of scenic views: hedonic assessment of private housing in Hong Kong', *Landscape and Urban Planning*, 91: 226–234.

Jim, C.Y. and Chen, W.Y. (2010) 'Habitat effect on vegetation ecology and occurrence on urban masonry walls', *Urban Forestry and Urban Greening*, 9: 169–178.

Jochner, S.C., Sparks, T.H., Estrella, N. and Menzel, A. (2012) 'The influence of altitude and urbanisation on trends and mean dates in phenology (1980–2009)', *International Journal of Biometeorology*, 56: 387–394.

Johnson, C.Y., Bowker, J.M., Bergstrom, J.C. and Cordell, H.K. (2004) 'Wilderness values in America: does immigrant status or ethnicity matter?', *Society and Natural Resources*, 17: 611–628.

Johnson, S.N., Elston, D.A. and Hartley, S.E. (2003) 'Influence of host plant heterogeneity on the distribution of a birch aphid', *Ecological Entomology*, 28: 533–541.

Jokimaki, J., Kaisanlahti-Jokimaki, M.L., Suhonen, J., Clergeau, P., Pautasso, M. and Fernandez-Juricic, E. (2011) 'Merging wildlife community ecology with animal behavioral ecology for a better urban landscape planning', *Landscape and Urban Planning*, 100: 383–385.

Jones, R.A. (2006) '*Brachinus sclopeta (Fabricius)* (Coleoptera) confirmed as a British species', *The Coleopterist*, 15: 29–33.

Jongman, R.H.G., Kulvik, M. and Kristiansen, I. (2004) 'European ecological networks and greenways', *Landscape and Urban Planning*, 68: 305–319.

Joppa, L.N., Roberts, D.L., Myers, N. and Pimm, S.L. (2011) 'Biodiversity hotspots house most undiscovered plant species', *Proceedings of the National Academy of Sciences of the USA*, 108: 13171–13176.

Jorgensen, A. and Tylecote, M. (2007) 'Ambivalent landscapes: wilderness in the urban interstices', *Landscape Research*, 32: 443–462.

Junker, B. and Buchecker, M. (2008) 'Aesthetic preferences versus ecological objectives in river restorations', *Landscape and Urban Planning*, 85: 141–154.

Kabat, T.J., Stewart, G.B. and Pullin, A.S. (2006) 'Are Japanese knotweed (*Fallopia japonica*) control and eradication interventions effective?', *Systematic Review*, No. 21. Collaboration for Environmental Evidence. Online. Available: http://www.environmentalevidence.org/Documents/Completed_Reviews/SR21.pdf (accessed 28 August 2012).

Kadas, G. (2006) 'Rare invertebrates colonizing green roofs in London', *Urban Habitats*, 4: 66–86.

Kalwij, J.M., Milton, S.J. and McGeoch, M.A. (2008) 'Road verges as invasion corridors? A spatial hierarchical test in an arid ecosystem', *Landscape Ecology*, 23: 439–451.

Kareiva, P., Tallis, H., Ricketts, T.H., Daily, G.C. and Polasky, S. (eds) (2011) *Natural Capital: Theory and Practice of Mapping Ecosystem Services*. Oxford: Oxford University Press.

Kasanko, M., Barredo, J.I., Lavalle, C., McCormick, N., Demicheli, L., Sagris, V. and Brezger, A. (2006) 'Are European cities becoming dispersed? A comparative analysis of 15 European urban areas', *Landscape and Urban Planning*, 77: 111–130.

Kattwinkel, M., Biedermann, R. and Kleyer, M. (2011) 'Temporary conservation for urban biodiversity', *Biological Conservation*, 144: 2335–2343.

Kaye, J.P., Burke, I.C., Mosier, A.R. and Guerschman, J.P. (2004) 'Methane and nitrous oxide fluxes from urban soils to the atmosphere', *Ecological Applications*, 14: 975–981.

Kaye, J.P., Groffman, P.M., Grimm, N.B., Baker, L.A. and Pouyat, R.V. (2006) 'A distinct urban biogeochemistry?', *Trends in Ecology and Evolution*, 21: 192–199.

Kaye, J.P., Majumdar, A., Gries, C., Buyantuycv, A., Grimm, N.B., Hope, D., Jenerette, G.D., Zhu, W.X. and Baker, L. (2008) 'Hierarchical Bayesian scaling of soil properties across urban, agricultural, and desert ecosystems', *Ecological Applications*, 18: 132–145.

Keefe, E.M. and Giuliano, W.M. (2004) 'Effects of forest structure on the distribution of southern flying squirrels (*Glaucomys volans*) in urban parks', *Urban Ecosystems*, 7: 55–64.

Kennedy, C., Pincetl, S. and Bunje, P. (2011) 'The study of urban metabolism and its applications to urban planning and design', *Environmental Pollution*, 159: 1965–1973.

Kiesecker, J.M., Copeland, H., Pocewicz, A. and McKenney, B. (2010) 'Development by design: blending landscape-level planning with the mitigation hierarchy', *Frontiers in Ecology and the Environment*, 8: 261–266.

Kim, K.C. and Byrne, L.B. (2006) 'Biodiversity loss and the taxonomic bottleneck: emerging biodiversity science', *Ecological Research*, 21: 794–810.

Kim, N.C. (2005) 'Ecological restoration and revegetation works in Korea', *Landscape and Ecological Engineering*, 1: 77–83.

Kim, S. (2007) 'Changes in the nature of urban spatial structure in the United States, 1890–2000', *Journal of Regional Science*, 47: 273–287.

Kinzig, A.P., Warren, P., Martin, C., Hope, D. and Katti, M. (2005) 'The effects of human socioeconomic status and cultural characteristics on urban patterns of biodiversity', *Ecology and Society*, 10: 23.

Klotz, S. (1990) 'Species/area and species/inhabitants relations in European cities', in H. Sukopp, S. Hegný and I. Kowarik (eds) *Urban Ecology*. The Hague: SPB Academic Publishing.

Klysik, K. and Fortuniak, K. (1999) 'Temporal and spatial characteristics of the urban heat island of Lodz, Poland', *Atmospheric Environment*, 33: 24–25.

Koch, F.H., Yemshanov, D., Colunga-Garcia, M., Magarey, R.D. and Smith, W.D. (2011) 'Potential establishment of alien-invasive forest insect species in the United States: where and how many?', *Biological Invasions*, 13: 969–985.

Köhler, M. (2008) 'Green facades: a view back and some visions', *Urban Ecosystems*, 11: 423–436.

Kondolf, G.M., Anderson, S.D., Storesund, R., Tompkins, M. and Atwood, P. (2011) 'Post-project appraisals of river restoration in advanced university instruction', *Restoration Ecology*, 19: 696–700.

Konijnendijk, C.C. (1999) 'Urban forestry policy-making: a comparative study of selected cities in Europe', *Arboricultural Journal*, 21: 1–15.

Kostylev, V.E., Erlandsson, J., Ming, M.Y. and Williams, G.A. (2005) 'The relative importance of habitat complexity and surface area in assessing biodiversity: fractal application on rocky shores', *Ecological Complexity*, 2: 272–286.

Kowarik, I. (1990) 'Some responses of flora and vegetation to urbanization in central Europe', in H. Sukopp, S. Hegný and I. Kowarik (eds) *Urban Ecology*. The Hague: SPB Academic Publishing.

Krasny, M.E. and Tidball, K.G. (2012) 'Civic ecology: a pathway for earth stewardship in cities', *Frontiers in Ecology and the Environment*, 10: 267–273.

Kremen, C. and Ostfeld, R.S. (2005) 'A call to ecologists: measuring, analyzing, and managing ecosystem services', *Frontiers in Ecology and Environment*, 3: 540–548.

Kühnert, C., Helbing, D. and West, G.B. (2006) 'Scaling laws in urban supply networks', *Physica A – Statistical Mechanics and Its Applications*, 363: 96–103.

Kupfer, J.A. (2012) 'Landscape ecology and biogeography: rethinking landscape metrics in a post-FRAGSTATS landscape', *Progress in Physical Geography*, 36: 400–420.

Lafortezza, R., Corry, R.C., Sanesi, G. and Brown, R.D. (2008) 'Visual preference and ecological assessments for designed alternative brownfield rehabilitations', *Journal of Environmental Management*, 89: 257–269.

Lake, P.S., Bond, N. and Reich, P. (2007) 'Linking ecological theory with stream restoration', *Freshwater Biology*, 52: 597–615.

LaPaix, R. and Freedman, B. (2010) 'Vegetation structure and composition within urban parks of Halifax Regional Municipality, Nova Scotia, Canada', *Landscape and Urban Planning*, 98: 124–135.

Larson, D.W., Matthes, U., Kelly, P.E., Lundholm, J.T. and Gerrath, J.A. (2004) *The Urban Cliff Revolution: New Findings on the Origins and Evolution of Human Habitats*. Markham, Canada: Fitzhenry and Whiteside.

Larson, M.G., Booth, D.B. and Morley, S.A. (2001) 'Effectiveness of large woody debris in stream rehabilitation projects in urban basins', *Ecological Engineering*, 18: 211–226.

Laurance, W.F. (2008) 'Theory meets reality: how habitat fragmentation research has transcended island biogeographic theory', *Biological Conservation*, 141: 1731–1744.

Leake, J.R., Adam-Bradford, A. and Rigby, J.E. (2009) 'Health benefits of "grow your own" food in urban areas: implications for contaminated land risk assessment and risk management?', *Environmental Health*, 8: S6.

Leichenko, R. (2011) 'Climate change and urban resilience', *Current Opinion in Environmental Sustainability*, 3: 164–168.

Leisnham, P. (2012) '*Aedes albopictus Skuse* (Asian tiger mosquito)', in R.A. Francis (ed.) *A Handbook of Global Freshwater Invasive Species*. Abingdon, Oxon: Earthscan.

Lerner, D.N. (2002) 'Identifying and quantifying urban recharge: a review', *Hydrogeology Journal*, 10: 143–152.

Levin, S.A. (1998) 'Ecosystems and the biosphere as complex adaptive systems', *Ecosystems*, 1: 431–436.

Lewis, D.B., Kaye, J.P., Gries, C., Kinzig, A.P. and Redman, C.L. (2006) 'Agrarian legacy in soil nutrient pools of urbanizing arid lands', *Global Change Biology*, 12: 703–709.

Li, F., Wang, R., Paulussen, J. and Liu, X. (2005) 'Comprehensive concept planning of urban greening based on ecological principles: a case study in Beijing, China', *Landscape and Urban Planning*, 72: 325–336.

Livesley, S.J., Dougherty, B.J., Smith, A.J., Navaud, D., Wylie, L.J. and Arndt, S.K. (2010) 'Soil–atmosphere exchange of carbon dioxide, methane and nitrous oxide in urban garden systems: impact of irrigation, fertiliser and mulch', *Urban Ecosystems*, 13: 273–293.

Lizée, M.H., Mauffrey, J.F., Tatoni, T. and Deschamps-Cottin, M. (2011) 'Monitoring urban environments on the basis of biological traits', *Ecological Indicators*, 11: 353–361.

Lobel, S., Snall, T. and Rydin, H. (2006) 'Metapopulation processes in epiphytes inferred from patterns of regional distribution and local abundance in fragmented forest landscapes', *Journal of Ecology*, 94: 856–868.

Loeb, R.E. (2006) 'A comparative flora of large urban parks: intraurban and interurban similarity in the megalopolis of the northeastern United States', *Journal of the Torrey Botanical Society*, 133: 601–625.

Longcore, T. and Rich, C. (2004) 'Ecological light pollution', *Frontiers in Ecology and the Environment*, 2: 191–198.

Lorimer, J. (2006) 'What about the nematodes? Taxonomic partialities in the scope of UK biodiversity conservation', *Social and Cultural Geography*, 7: 539–558.

Lovell, S.T. and Johnston, D.M. (2009) 'Designing landscapes for performance based on emerging principles in landscape ecology', *Ecology and Society*, 14: 44.

LTER Network (2011a) 'LTER Network Mission and Vision Statements'. Online. Available: http://www.lternet.edu/mission/ (accessed 25 July 2012).

LTER Network (2011b) 'Baltimore Ecosystem Study'. Online. Available: http://www.lternet.edu/sites/bes/ (accessed 25 July 2012).

LTER Network (2011c) 'Central Arizona-Phoenix LTER'. Online. Available: http://www.lternet.edu/sites/cap/ (accessed 25 July 2012).

Lubbe, C.S., Siebert, S.J. and Cilliers, S.S. (2010) 'Political legacy of South Africa affects the plant diversity patterns of urban domestic gardens along a socio-economic gradient', *Scientific Research and Essays*, 5: 2900–2910.

Luck, M. and Wu, J.G. (2002) 'A gradient analysis of urban landscape pattern: a case study from the Phoenix metropolitan region, Arizona, USA', *Landscape Ecology*, 17: 327–339.

Luck, M.A., Jenerette, G.D., Wu, J.G. and Grimm, N.B. (2001) 'The urban funnel model and the spatially heterogeneous ecological footprint', *Ecosystems*, 4: 782–796.

Lugo, A.E. (2010) 'Let's not forget the biodiversity of the cities', *Biotropica*, 42: 576–577.

Lundholm, J.T. (2011) 'Vegetation of urban hard surfaces', in J. Niemelä, J.H. Breuste, G. Guntenspergen, N.E. McIntyre, T. Elmqvist and P. James (eds) *Urban Ecology: Patterns, Processes, and Applications*. Oxford: Oxford University Press.

Lundholm, J.T. and Richardson, R.J. (2010) 'Habitat analogues for reconciliation ecology in urban and industrial environments', *Journal of Applied Ecology*, 47: 966–975.

Lundy, L. and Wade, R. (2011) 'Integrating sciences to sustain urban ecosystem services', *Progress in Physical Geography*, 35: 653–669.

Luniak, M. (2004) 'Synurbization: adaptation of animal wildlife to urban development', in W.W. Shaw, L.K. Harris and L. VanDruff (eds) *Proceedings of the 4th International Symposium on Urban Wildlife Conservation*. Tucson: University of Arizona.

Lutz, W. and Samir, K.C. (2010) 'Dimensions of global population projections: what do we know about future population trends and structures?', *Philosophical Transactions of the Royal Society B*, 365: 2779–2791.

Lyytimäki, J. and Sipilä, M. (2009) 'Hopping on one leg: the challenge of ecosystem disservices for urban green management', *Urban Forestry and Urban Greening*, 8: 309–315.

Macadam, C.R. and Bairner, S.Z. (2012) 'Urban biodiversity: Successes and challenges: Brownfields: oases of urban biodiversity', *The Glasgow Naturalist*, 25: 1–4.

MacArthur, R.H. and Wilson, E.O. (1967) *The Theory of Island Biogeography*. Princeton, NJ: Princeton University Press.

MacGregor-Fors, I. (2011) 'Misconceptions or misunderstandings? On the standardization of basic terms and definitions in urban ecology', *Landscape and Urban Planning*, 100: 347–349.

MacGregor-Fors, I. and Ortega-Alvarez, R. (2011) 'Fading from the forest: bird community shifts related to urban park site-specific and landscape traits', *Urban Forestry and Urban Greening*, 10: 239–246.

Mack, R.N. (2003) 'Global plant dispersal, naturalization, and invasion: pathways, modes and circumstances'. in G.M. Ruiz and J.T. Carlton (eds) *Invasive Species: Vectors and Management Strategies*. Washington, DC: Island Press.

Magle, S.B., Theobald, D.M. and Crooks, K.R. (2009) 'A comparison of metrics predicting landscape connectivity for a highly interactive species along an urban gradient in Colorado, USA', *Landscape Ecology*, 24: 267–280.

Magle, S.B., Reyes, P., Zhu, J. and Crooks, K.R. (2010) 'Extirpation, colonization, and habitat dynamics of a keystone species along an urban gradient', *Biological Conservation*, 143: 2146–2155.

Mahan, C.G. and O'Connell, T.J. (2005) 'Small mammal use of suburban and urban parks in central Pennsylvania', *Northeastern Naturalist*, 12: 307–314.

Marcuse, P. and van Kempen, R. (2002) *Of States and Cities: The Partitioning of Urban Space*. London: Oxford University Press.

Marshall, M. (2012) 'Will there be any nature left?', *New Scientist*, 213: 43–44.

Martinez, M.L., Intralawan, A., Vazquez, G., Perez-Maqueo, O., Sutton, P. and Landgrave, R. (2007) 'The coasts of our world: ecological, economic and social importance', *Ecological Economics*, 63: 254–272.

Marzluff, J.M. and Ewing, K. (2001) 'Restoration of fragmented landscapes for the conservation of birds: a general framework and specific recommendations for urbanizing landscapes', *Restoration Ecology*, 9: 280–292.

Masek, J.G., Lindsay, F.E. and Goward, S.N. (2000) 'Dynamics of urban growth in the Washington DC metropolitan area, 1973–1996, from Landsat observations', *International Journal of Remote Sensing*, 21: 3473–3486.

Masi, E., Pino, F.A., Santos, M.D.S., Genehr, L., Albuquerque, J.O.M., Bancher, A.M. and Alves, J.C.M. (2010) 'Socioeconomic and environmental risk factors for urban rodent infestation in Sao Paulo, Brazil', *Journal of Pest Science*, 83: 228–238.

Mathieu, R., Freeman, C. and Aryal, J. (2007) 'Mapping private gardens in urban areas using object-oriented techniques and very high-resolution satellite imagery', *Landscape and Urban Planning*, 81: 179–192.

Matos, D.M.S., Santos, C.J.F. and Chevalier, D.R. (2003) 'Fire and restoration of the largest urban forest of the world in Rio de Janeiro City, Brazil', *Urban Ecosystems*, 6: 151–161.

Matteson, K.C., Taron, D.J. and Minor, E.S. (2012) 'Assessing citizen contributions to butterfly monitoring in two large cities', *Conservation Biology*, 26: 557–564,

May, R.M. (2010) 'Ecological science and tomorrow's world', *Philosophical Transactions of the Royal Society B*, 365: 41–47.

McConnachie, M. and Shackleton, C.M. (2010) 'Public green space inequality in small towns in South Africa', *Habitat International*, 34: 244–248.

McCook, L.J. (1994) 'Understanding ecological community succession: causal models and theories, a review', *Vegetatio*, 110: 115–147.

McDonnell, M.J., Hahs, A.K. and Pickett, S.T.A. (2012) 'Exposing an urban ecology straw man: critique of Ramalho and Hobbs', *Trends in Ecology and Evolution*, 27: 255–256.

McGee, T. (1989) 'Urbanisasi or kotadesasi? Evolving patterns of urbanization in Asia', in F.J. Costa, A.K. Dutt, L.C.J. Ma and A.G. Noble (eds) *Urbanization in Asia*. Honolulu: University of Hawaii Press.

McGeoch, M.A., Butchart, S.H.M., Spear, D., Marais, E., Kleynhans, E.J., Symes, A., Chanson, J. and Hoffmann, M. (2010) 'Global indicators of biological invasion: species numbers, biodiversity impact and policy responses', *Diversity and Distributions*, 16: 95–108.

McIntyre, N.E. (2000) 'Ecology of urban arthropods: a review and a call to action', *Annals of the Entomological Society of America*, 93: 825–835.

McIntyre, S. and Hobbs, R. (1999) 'A framework for conceptualizing human effects on land-scapes and its relevance to management and research models', *Conservation Biology*, 13: 1282–1292.

McKinney, M.L. (2008) 'Effects of urbanization on species richness: a review of plants and animals', *Urban Ecosystems*, 11: 161–176.

McLellan, S.L., Huse, S.M., Mueller-Spitz, S.R., Andreishcheva, E.N. and Sogin, M.L. (2010) 'Diversity and population structure of sewage-derived microorganisms in wastewater treat-ment plant influent', *Environmental Microbiology*, 12: 378–392.

Megson, D., Dack, S. and Moore, M. (2011) 'Limitations of the CLEA model when assessing human health risks from dioxins and furans in soil at an allotments site in Rochdale, NW England', *Journal of Environmental Modelling*, 13: 1983–1990.

Meiner, S.J. and Pickett, S.T.A. (2011) 'Succession', in D. Simberloff and M. Rejmanek (eds) *Encyclopedia of Biological Invasions*. Berkeley: University of California Press.

Metz, S. (2000) *Armed Conflict in the 21st Century: The Information Revolution and Post-Modern Warfare*. Carlisle, PA: Strategic Studies Institute, United States Army War College.

Meurk, C.D. (2010) 'Recombinant ecology of urban areas: characterisation, context and crea-tivity', in I. Douglas, D. Goode, M. Houck and R. Wang (eds) *The Routledge Handbook of Urban Ecology*, London: Routledge.

Miller, J.R. (2006) 'Restoration, reconciliation, and reconnecting with nature nearby', *Biological Conservation*, 127: 356–361.

Mills, L.S., Soulé, M.E. and Doak, D.F. (1993) 'The keystone-species concepts in ecology and conservation', *BioScience*, 43: 219–224.

Miltner, R.J., White, D. and Yoder, C. (2004) 'The biotic integrity of streams in urban and suburbanizing landscapes', *Landscape and Urban Planning*, 69: 87–100.

Milton, S.J. (2003) ' "Emerging ecosystems" – a washing-stone for ecologists, economists and sociologists?', *South African Journal of Science*, 99: 404–406.

Mohiuddin, K.M., Zakir, H.M., Otomo, K., Sharmin, S. and Shikazono, N. (2010) 'Geochemical distribution of trace metal pollutants in water and sediments of downstream of an urban river', *International Journal of Environmental Science and Technology*, 7: 17–28.

Molina, M.J. and Molina, L.T. (2004) 'Megacities and atmospheric pollution', *Journal of the Air and Waste Management Association*, 54: 644–680.

Møller, A.P. (2009) 'Successful city dwellers: a comparative study of the ecological characteris-tics of urban birds in the Western Palearctic', *Oecologia*, 159: 849–858.

Mooney, H.A. (2010) 'The ecosystem-service chain and the biological diversity crisis', *Philosophical Transactions of the Royal Society B*, 365: 31–39.

Mörtberg, U.M. (2001) 'Resident bird species in urban forest remnants: landscape and habitat perspectives', *Landscape Ecology*, 16: 193–203.

Mörtberg, U.M., Balfors, B. and Knol, W.C. (2007) 'Landscape ecological assessment: a tool for integrating biodiversity issues in strategic environmental assessment and planning', *Journal of Environmental Management*, 82: 457–470.

Motard, E., Muratet, A., Clair-Maczulajtys, D. and Machon, N. (2011) 'Does the invasive spe-cies *Ailanthus altissima* threaten floristic diversity of temperate pen-urban forests?', *Comptes Rendus Biologies*, 334: 872–879.

Munshi-South, J. (2012) 'Urban landscape genetics: canopy cover predicts gene flow between white-footed mouse (*Peromyscus leucopus*) populations in New York City', *Molecular Ecology*, 21: 1360–1378.

Muratet, A., Machon, N., Jiguet, F., Moret, J. and Porcher, E. (2007) 'The role of urban structures in the distribution of wasteland flora in the greater Paris area, France', *Ecosystems*, 10: 661–671.

Murray, P., Ge, Y. and Hendershot, W.H. (2000) 'Evaluating three trace metal contaminated sites: a field and laboratory investigation', *Environmental Pollution*, 107: 127–135.

Mustafa, A., Scholz, M., Harrington, R. and Carroll, P. (2009) 'Long-term performance of a representative integrated constructed wetland treating farmyard runoff', *Ecological Engineering*, 35: 779–790.

Nabulo, G., Young, S.D. and Black, C.R. (2010) 'Assessing risk to human health from tropical leafy vegetables grown on contaminated urban soils', *Science of the Total Environment*, 408: 5338–5351.

Nahmani, J. and Lavelle, P. (2002) 'Effects of heavy metal pollution on soil macrofauna in a grassland of Northern France', *European Journal of Soil Biology*, 38: 297–300.

Natuhara, Y. and Imai, C. (1999) 'Prediction of species richness of breeding birds by landscape-level factors of urban woods in Osaka Prefecture, Japan', *Biodiversity and Conservation*, 8: 239–253.

Nedelcheva, A. and Vasileva, A. (2009) 'Vascular plants from the old walls in Kystendil (Southwestern Bulgaria)', *Biotechnology and Biotechnological Equipment*, 23: 154–157.

Nemec, K.T., Allen, C.R., Alai, A., Clements, G., Kessler, A.C., Kinsell, T., Major, A. and Stephen, B.J. (2011) 'Woody invasions of urban trails and the changing face of urban forests in the Great Plains, USA', *American Midland Naturalist*, 165: 241–256.

Newing, H. (2010) 'Interdisciplinary training in environmental conservation: definitions, progress and future directions', *Environmental Conservation*, 37: 410–418.

Nicholls, R.J. (1995) 'Coastal megacities and climate change', *GeoJournal*, 37: 369–379.

Niemelä, J., Kotze, D.J., Venn, S., Penev, L., Stoyanov, I., Spence, J., Hartley, D. and de Oca, E.M. (2002) 'Carabid beetle assemblages (Coleoptera, Carabidae) across urban–rural gradients: an international comparison', *Landscape Ecology*, 17: 387–401.

Niu, H., Clark, C., Zhou, J. and Adriaens, P. (2010) 'Scaling of economic benefits from green roof implementation in Washington, DC', *Environmental Science and Technology*, 44: 4302–4308.

Novy, A., Redak, V., Jager, J. and Hamedinger, A. (2001) 'The end of red Vienna: recent ruptures and continuities in urban governance', *European Urban and Regional Studies*, 8: 131–144.

Oberndorfer, E., Lundholm, J., Bass, B., Coffman, R.R., Doshi, H., Dunnett, N., Gaffin, S., Köhler, M., Liu, K.K.Y. and Rowe, B. (2007) 'Green roofs as urban ecosystems: ecological structures, functions and services', *BioScience*, 57: 823–833.

Odum, E.P. (1997) *Ecology: A Bridge Between Science and Society*. Sunderland, MA: Sinauer.

Odum, H.T. (1983) *Systems Ecology: An Introduction*. New York: John Wiley & Sons.

Odum, H.T. (1988) 'Self-organisation, transformity, and information', *Science*, 242: 1132–1139.

Okabe, S., Odagiri, M., Ito, T. and Satoh, H. (2007) 'Succession of sulfur-oxidizing bacteria in the microbial community on corroding concrete in sewer systems', *Applied and Environmental Microbiology*, 73: 971–980.

Oliver, A.J., Hong-Wa, C., Devonshire, J., Olea, K.R., Rivas, G.F. and Gahl, M.K. (2011) 'Avifauna richness enhanced in large, isolated urban parks', *Landscape and Urban Planning*, 102: 215–225.

Olly, L.M., Bates, A.J., Sadler, J.P. and Mackay, R. (2011) 'An initial experimental assessment of the influence of substrate depth on floral assemblage for extensive green roofs', *Urban Forestry and Urban Greening*, 10: 311–316.

Olympic Delivery Authority (2008) *Olympic Park Biodiversity Action Plan*. London: Olympic Delivery Authority. Online. Available: http://www.london2012.com/mm%5CDocument% 5CPublications%5CSustainability%5C01%5C24%5C08%5C11%5Colympic-park-bio diversity-action-plan.pdf (accessed 10 August 2012).

Ooi, G.L. (2011) 'The role of the state in nature conservation in Singapore', *Society and Natural Resources: An International Journal*, 15: 455–460.

Orłowski, G. (2008) 'Roadside hedgerows and trees as factors increasing road mortality of birds: implications for management of roadside vegetation in rural landscapes', *Landscape and Urban Planning*, 86: 153–161.

Ourso, R.T. and Frenzel, S.A. (2003) 'Identification of linear and threshold responses in streams along a gradient of urbanization in Anchorage, Alaska', *Hydrobiologia*, 501: 117–131.

Page, D., Dillon, P., Vanderzalm, J., Toze, S., Sidhu, J., Barry, K., Levett, K., Kremer, S. and Regel, R. (2010) 'Risk assessment of aquifer storage transfer and recovery with urban storm-water for producing water of a potable quality', *Journal of Environmental Quality*, 39: 2029–2039.

Papanastasiou, D.K. and Melas, D. (2009) 'Climatology and impact on air quality of sea breeze in an urban coastal environment', *International Journal of Climatology*, 29: 305–315.

Parris, K.M. (2006) 'Urban amphibian assemblages as metacommunities', *Journal of Animal Ecology*, 75: 757–764.

Pauchard, A., Aguayo, M., Pena, E. and Urrutia, R. (2006) 'Multiple effects of urbanization on the biodiversity of developing countries: the case of a fast-growing metropolitan area (Concepcion, Chile)', *Biological Conservation*, 127: 272–281.

Paul, M.J. and Meyer, J.L. (2001) 'Streams in the urban landscape', *Annual Review of Ecology and Systematics*, 32: 333–365.

Pauleit, S., Jones, N., Garcia-Martin, G., Garcia-Valdecantos, J-L., Riviere, L-M., Vidal-Beaudet, L., Bodson, M. and Randrup, T.B. (2002) 'Tree establishment practice in towns and cities: results from a European survey', *Urban Forestry and Urban Greening*, 1: 83–96.

Pauleit, S., Ennos, R. and Golding, Y. (2005) 'Modeling the environmental impacts of urban land use and land cover change: a study in Merseyside, UK', *Landscape and Urban Planning*, 71: 295–310.

Pautasso, M. (2007) 'Scale dependence of the correlation between human population presence and vertebrate and plant species richness', *Ecology Letters*, 10: 16–24.

Pavao-Zuckerman, M.A. (2008) 'The nature of urban soils and their role in ecological restoration in cities', *Restoration Ecology*, 16: 642–649.

Pavao-Zuckerman, M.A. and Coleman, D.C. (2007) 'Urbanization alters the functional composition, but not taxonomic diversity, of the soil nematode community', *Applied Soil Ecology*, 35: 329–339.

Peffy, T. and Nawaz, R. (2008) 'An investigation into the extent and impacts of hard surfacing of domestic gardens in an area of Leeds, United Kingdom', *Landscape and Urban Planning*, 86: 1–13.

Peiser, S.B. (1989) 'Density and urban sprawl', *Land Economics*, 65: 193–204.

Penone, C., Machon, N., Julliard, R. and Le Viol, I. (2012) 'Do railway edges provide functional connectivity for plant communities in an urban context?', *Biological Conservation*, 148: 126–133.

Pergl, J., Perglová, I. and Pyšek, P. (2012) '*Heracleum mantegazzianum* Sommier and Levier (giant hogweed)', in R.A. Francis (ed.) *A Handbook of Global Freshwater Invasive Species*. Abingdon, Oxon: Earthscan.

Petts, G.E., Heathcote, J. and Martin, D. (2002) *Urban Rivers: Our Inheritance and Future*. London: IWA Publishing.

Pickett, S.T.A. and Cadenasso, M.L. (2002) 'The ecosystem as a multidimensional concept: meaning, model, and metaphor', *Ecosystems*, 5: 1–10.

Pickett, S.T.A. and Cadenasso, M.L. (2006) 'Advancing urban ecological studies: frameworks, concepts, and results from the Baltimore Ecosystem Study', *Austral Ecology*, 31: 114–125.

Pickett, S.T.A. and Grove, J.M. (2009) 'Urban ecosystems: what would Tansley do?', *Urban Ecosystems*, 12: 1–8.

Pickett, S.T.A., Cadenasso, M.L., Grove, J.M., Nilon, C.H., Pouyat, R.V., Zipperer, W.C. and Costanza, R. (2001) 'Urban ecological systems: linking terrestrial ecological, physical, and socioeconomic components of metropolitan areas', *Annual Review of Ecology and Systematics*, 32: 127–157.

Pickett, S.T.A., Cadenasso, M.L., Grove, J.M., Groffman, P.M., Band, L.E., Boone, C.G., Burch, W.R., Grimmond, C.S.B., Hom, J., Jenkins, J.C., Law, N.L., Nilon, C.H., Pouyat, R.V., Szlavecz, K., Warren, P.S. and Wilson, M.A. (2008) 'Beyond urban legends: an emerging framework of urban ecology, as illustrated by the Baltimore Ecosystem Study', *BioScience*, 58: 139–150.

Pickett, S.T.A., Cadenasso, M.L., Grove, J.M., Boone, C.G., Groffman, P.M., Irwin, E., Kaushal, S.S., Marshall, V., McGrath, B.P., Nilon, C.H., Pouyat, R.V., Szlavecz, K., Troy, A. and Warren, P. (2011) 'Urban ecological systems: Scientific foundations and a decade of progress', *Journal of Environmental Management*, 92: 331–362.

Pimentel, D., Zuniga, R. and Morrison, D. (2005) 'Update on the environmental and economic costs associated with alien-invasive species in the United States', *Ecological Economics*, 52: 273–288.

Pino, J. and Marull, J. (2012) 'Ecological networks: are they enough for connectivity conservation? A case study in the Barcelona Metropolitan Region (NE Spain)', *Land Use Policy*, 29: 684–690.

Poor, N.D. (2010) 'Effect of lake management efforts on the trophic state of a subtropical shallow lake in Lakeland, Florida, USA', *Water, Air and Soil Pollution*, 207: 333–347.

Porter, E.E., Forschner, B.R. and Blair, R.B. (2001) 'Woody vegetation and canopy fragmentation along a forest-to-urban gradient', *Urban Ecosystems*, 5: 131–151.

Potter, M.F. (2005) 'A bed bug state of mind: emerging issues in bed bug management', *Pest Control Technology*, 33: 82–85.

Potter, M.F., Rosenberg, B. and Henriksen, M (2010) 'Bugs without borders: defining the global bed bug resurgence', *Pest World*, September/October: 8–20.

Pouyat, R.V., Yesilonis, I.D. and Nowak, D.J. (2006) 'Carbon storage by urban soils in the United States', *Journal of Environmental Quality*, 35: 1566–1575.

Pouyat, R.V., Szlavecz, K., Yesilonis, I.D., Groffman, P.M. and Schwarz, K. (2010) 'Chemical, physical, and biological characteristics of urban soils', in J. Aitkenhead-Peterson and A. Volder (eds) *Urban Ecosystem Ecology: Agronomy Monograph 55*. Madison, WI: American Society of Agronomy, Inc., Crop Science Society of America, Inc., Soil Science Society of America, Inc.

Prevedello, J.A. and Vieira, M.V. (2010) 'Does the type of matrix matter? A quantitative review of the evidence', *Biodiversity and Conservation*, 19: 1205–1223.

Pullin, A.S. and Knight, T.M. (2009) 'Doing more good than harm: building an evidence-base for conservation and environmental management', *Biological Conservation*, 142: 931–934.

Purcell, A.H., Bressler, D.W., Paul, M.J., Barbour, M.T., Rankin, E.T., Carter, J.L. and Resh, V.H. (2009) 'Assessment tools for urban catchments: developing biological indicators based on benthic macroinvertebrates', *Journal of the American Water Resources Association*, 45: 306–319.

Quigley, M.F. (2004) 'Street trees and rural conspecifics: will long-lived trees reach full size in urban conditions?', *Urban Ecosystems*, 7: 29–39.

Rackham, O. (2000) *The History of the Countryside*. London: Phoenix Press.

Radeloff, V.C., Hammer, R.B. and Stewart, S.I. (2005) 'Rural and suburban sprawl in the US Midwest from 1940 to 2000 and its relation to forest fragmentation', *Conservation Biology*, 19: 793–805.

Ramalho, C.E. and Hobbs, R.J. (2012) 'Time for a change: dynamic urban ecology', *Trends in Ecology and Evolution*, 27: 179–188.

Ranta, P., Tanskanen, A., Niemelä, J. and Kurtto, A. (1999) 'Selection of islands for conservation in the urban archipelago of Helsinki, Finland', *Conservation Biology*, 13: 1293–1300.

Rebele, F. (1994) 'Urban ecology and special features of urban ecosystems', *Global Ecology and Biogeography Letters*, 4: 173–187.

Reckien, D. and Martinez-Fernandez, C. (2011) 'Why do cities shrink?', *European Planning Studies*, 19: 1375–1397.

Redman, C.L., Grove, J.M. and Kuby, L.H. (2004) 'Integrating social science into the long-term ecological research (LTER) network: social dimensions of ecological change and ecological dimensions of social change', *Ecosystems*, 7: 161–171.

Reid, A.M. and Hochuli, D.F. (2007) 'Grassland invertebrate assemblages in managed landscapes: effect of host plant and microhabitat architecture', *Austral Ecology*, 32: 708–718.

Reinhardt, K., Harder, A., Holland, S., Hooper, J. and Leake-Lyall, C. (2008) 'Who knows the bed bug? Knowledge of adult bed bug appearance increases with people's age in three counties of Great Britain', *Journal of Medical Entomology*, 45: 956–958.

Reuben, S., Chua, C.L.N., Fam, K.D., Thian, Z.Y.A., Kang, M.K. and Swarup, S. (2012) 'Bacterial diversity on different surfaces in urban freshwater', *Water Science and Technology*, 65: 1869–1874.

Reyes-Paecke, S. and Meza, L. (2011) 'Residential gardens of Santiago, Chile: extent, distribution and vegetation cover', *Revista Chilena de Historia Natural*, 84: 581–592.

Ricciardi, A. and Cohen, J. (2007) 'The invasiveness of an introduced species does not predict its impact', *Biological Invasions*, 9: 309–315.

Richardson, D.M. and Pyšek, P. (2006) 'Plant invasions: merging the concepts of species invasiveness and community invasibility', *Progress in Physical Geography*, 30: 409–431.

Richmond-Bryant, J., Isukapalli, S.S. and Vallero, D.A. (2011) 'Air pollutant retention within a complex of urban street canyons', *Atmospheric Environment*, 45: 7612–7618.

Rietkerk, M., van, de and Koppel, J. (2008) 'Regular pattern formation in real ecosystems', *Trends in Ecology and Evolution*, 23: 169–175.

Riley, S.P.D., Busteed, G.T., Kats, L.B., Vandergon, T.L., Lee, L.F.S., Dagit, R.G., Kerby, J.L., Fisher, R.N. and Sauvajot, R.M. (2005) 'Effects of urbanization on the distribution and abundance of amphibians and invasive species in Southern California streams', *Conservation Biology*, 19: 1894–1907.

Robbins, P. and Birkenholtz, T. (2003) 'Turfgrass revolution: measuring the expansion of the American lawn', *Land Use Policy*, 20: 181–194.

Robinson, L., Newell, J.P. and Marzluff, J.A. (2005) 'Twenty-five years of sprawl in the Seattle region: growth management responses and implications for conservation', *Landscape and Urban Planning*, 71: 51–72.

Rock, P. (2005) 'Urban gulls: problems and solutions', *British Birds*, 98: 338–355.

Rodwell, J.S., Pigott, C.D., Ratcliffe, D.A., Malloch, A.J.C., Birks, H.J.B. and Proctor, M.C.F. (2000) *British Plant Communities Volume 5: Maritime Communities and Vegetation of Open Habitats*. Cambridge: Cambridge University Press.

Roedenbeck, I.A., Fahrig, L., Findlay, C.S., Houlahan, J.E., Jaeger, J.A.G., Klar, N., Kramer-Schadt, S. and van der Grift, E.A. (2007) 'The Rauischholzhausen Agenda for road ecology', *Ecology and Society*, 12: 11.

Roehr, D. and Kong, Y.W. (2010) 'Runoff reduction effects of green roofs in Vancouver, BC, Kelowna, BC, and Shanghai, PR China', *Canadian Water Resources Journal*, 35: 53–67.

Rosenzweig, M.L. (2003) 'Reconciliation ecology and the future of species diversity', *Oryx*, 37: 194–205.

Rowe, D.B., Getter, K.L. and Durhman, A.K. (2012) 'Effect of green roof media depth on Crassulacean plant succession over seven years', *Landscape and Urban Planning*, 104: 310–319.

Royal Institute of Chartered Surveyors (RICS) (2011) *Japanese Knotweed and Residential Property*. Consultation Draft November 2011. Online. Available: https://consultations. rics.org/gf2.ti/f/275138/6179845.1/PDF/-/Japanese%20Knotweed%20and%20residential %20property.pdf (accessed 28 August 2012).

Saarikivi, J., Idstrom, L., Venn, S., Niemelä, J. and Kotze, D.J. (2010) 'Carabid beetle assemblages associated with urban golf courses in the greater Helsinki area', *European Journal of Entomology*, 107: 553–561.

Saarinen, K., Valtonen, A., Jantunen, J. and Saarnio, S. (2005) 'Butterflies and diurnal moths along road verges: does road type affect diversity and abundance?', *Biological Conservation*, 123: 403–412.

Sadler, J., Bates, A., Hale, J. and James, P. (2010) 'Brining cities alive: the importance of urban green spaces for people and biodiversity', in K.J. Gaston (ed.) *Urban Ecology*. Cambridge: Cambridge University Press.

Sæbo, A., Benedikz, T. and Randrup, T.B. (2003) 'Selection of trees for urban forestry in the Nordic countries', *Urban Forestry and Urban Greening*, 2: 101–114.

Sandström, U.G., Angelstam, P. and Mikusinski, G. (2006) 'Ecological diversity of birds in relation to the structure of urban green space', *Landscape and Urban Planning*, 77: 39–53.

Santorufo, L., Van Gestel, C.A.M., Rocco, A. and Maisto, G. (2012) 'Soil invertebrates as bioindicators of urban soil quality', *Environmental Pollution*, 161: 57–63.

Säumel, I. and Kowarik, I. (2010) 'Urban rivers as dispersal corridors for primarily wind-dispersed invasive tree species', *Landscape and Urban Planning*, 94: 244–249.

Savard, J.P.L., Clergeau, P. and Mennechez, G. (2000) 'Biodiversity concepts and urban ecosystems', *Landscape and Urban Planning*, 48: 131–142.

Schadek, U., Strauss, B., Biedermann, R. and Kleyer, M. (2009) 'Plant species richness, vegetation structure and soil resources of urban brownfield sites linked to successional age', *Urban Ecosystems*, 12: 115–126.

Scharenbroch, B.C., Lloyd, J.E. and Johnson-Maynard, J.L. (2005) 'Distinguishing urban soils with physical, chemical, and biological properties', *Pedobiologia*, 49: 283–296.

Schenck, D.A. (2003–4) 'The next step for brownfields: government reinsurance of environmental "cleanup" policies', *Connecticut Insurance Law Journal*, 10: 401.

Schlünzen, K.H., Hoffmann, P., Rosenhagen, G. and Riecke, W. (2010) 'Long-term changes and regional differences in temperature and precipitation in the metropolitan area of Hamburg', *International Journal of Climatology*, 30: 1121–1136.

Schneider, A. and Woodcock, C.E. (2008) 'Compact, dispersed, fragmented, extensive? A comparison of urban growth in twenty-five global cities using remotely sensed data, pattern metrics and census information', *Urban Studies*, 45: 659–692.

Schneider, A., Friedl, M.A. and Potere, D. (2009) 'A new map of global urban extent from MODIS satellite data', *Environmental Research Letters*, 4: 044003.

Schrader, S. and Böning, M. (2006) 'Soil formation on green roofs and its contribution to urban biodiversity with emphasis on Collembolans', *Pedobiologia*, 50: 347–356.

Schwarz, N. (2010) 'Urban form revisited: selecting indicators for characterising European cities', *Landscape and Urban Planning*, 96: 29–47.

Segal, S. (1969) *Ecological Notes on Wall Vegetation*. The Hague, Netherlands: Dr. W. Junk N.V.

Šerá, B. (2008) 'Road vegetation in Central Europe: an example from the Czech Republic', *Biologia*, 63: 1085–1088.

Shanahan, D.F., Miller, C., Possingham, H.P. and Fuller, R.A. (2011) 'The influence of patch area and connectivity on avian communities in urban revegetation', *Biological Conservation*, 144: 722–729.

Shi, Y.Q., Sun, X., Zhu, X.D., Li, Y.F. and Mei, L.Y. (2012) 'Characterizing growth types and analyzing growth density distribution in response to urban growth patterns in peri-urban areas of Lianyungang City', *Landscape and Urban Planning*, 105: 425–433.

Shimwell, D.W. (2009) 'Studies in the floristic diversity of Durham walls, 1958–2008', *Watsonia*, 27: 323–338.

Shiode, N. (2001) '3D urban models: recent developments in the digital modelling of urban environments in three-dimensions', *GeoJournal*, 52: 263–269.

Shmida, A. and Ellner, S. (1984) 'Coexistence of plant species with similar niches', *Vegetatio*, 58: 29–55.

Shochat, E., Warren, P.S., Faeth, S.H., McIntyre, N.E. and Hope, D. (2006) 'From patterns to emerging processes in mechanistic urban ecology', *Trends in Ecology and Evolution*, 21: 186–191.

Sileshi, G.W., Kuntashula, E., Matakala, P. and Nkunika, P.O. (2008) 'Farmers' perceptions of tree mortality, pests and pest management practices in agroforestry in Malawi, Mozambique and Zambia', *Agroforestry Systems*, 72: 87–101.

Silvertown, J. (2009) 'A new dawn for citizen science', *Trends in Ecology and Evolution*, 24: 467–471.

Simberloff, D. (2008) 'We can eliminate invasions or live with them: successful management projects', *Biological Invasions*, 11: 149–157.

Sjöman, H., Ostberg, J. and Bühler, O. (2012) 'Diversity and distribution of the urban tree population in ten major Nordic cities', *Urban Forestry and Urban Greening*, 11: 31–39.

Small, E., Sadler, J.P. and Telfer, M. (2006) 'Do landscape factors affect brownfield carabid assemblages?', *Science of the Total Environment*, 360: 205–222.

Smallbone, L.T., Luck, G.W. and Wassens, S. (2011) 'Anuran species in urban landscapes: relationships with biophysical, built environment and socio-economic factors', *Landscape and Urban Planning*, 101: 43–51.

Smetak, K.M., Johnsn-Maynard, J.L. and Lloyd, J.E. (2007) 'Earthworm population density and diversity in different-aged urban systems', *Applied Soil Ecology*, 37: 161–168.

Smith, D.A. and Gehrt, S.D. (2010) 'Bat response to woodland restoration within urban forest fragments', *Restoration Ecology*, 18: 914–923.

Smith, R.M., Gaston, K.J., Warren, P.H. and Thompson, K. (2005) 'Urban domestic gardens (V): relationships between landcover composition, housing and landscape', *Landscape Ecology*, 20: 235–253.

Smith, R.M., Thompson, K., Hodgson, J.G., Warren, P.H. and Gaston, K.J. (2006a) 'Urban domestic gardens (IX): composition and richness of the vascular plant flora, and implications for native biodiversity', *Biological Conservation*, 129: 312–322.

Smith, R.M., Warren, P.H., Thompson, K. and Gaston, K.J. (2006b) 'Urban domestic gardens (VI): environmental correlates of invertebrate species richness', *Biodiversity and Conservation*, 15: 2415–2438.

Soh, M.C.K., Sodhi, N.S., Seoh, R.K.H. and Brook, B.W. (2002) 'Nest site selection of the house crow (*Corvus splendens*), an urban invasive bird species in Singapore and implications for its management', *Landscape and Urban Planning*, 59: 217–226.

Solarz, K., Senezuk, L., Maniurka, H., Cichecka, E. and Peszke, M. (2007) 'Comparisons of the allergenic mite prevalence in dwellings and certain outdoor environments of the Upper Silesia (southwest Poland)', *International Journal of Hygiene and Environmental Health*, 210: 715–724.

Sorace, A. (2001) 'Value to wildlife of urban-agricultural parks: a case study from Rome urban area', *Environmental Management*, 28: 547–560.

Soulé, M.E. (1990) 'The onslaught of alien species, and other challenges in the coming decades', *Conservation Biology*, 4: 233–239.

Spiller, D.A. and Schoener, T.W. (2009) 'Species–area relationship', in R.G. Gillespie and D.A. Clague (eds) *Encyclopedia of Islands*. Berkeley: University of California Press.

Stankowski, S.J. (1972) 'Population density as an indirect indicator of urban and suburban land-surface modifications', *Geological Survey Professional Paper*, 800-B: B219–B224.

Statistics Norway (2011) 'Population and land area in urban settlements, 1 January 2011'. Online. Available: http://www.ssb.no/beftett_en/ (accessed 22 August 2012).

Steinberg, D.A., Pouyat, R.V., Parmelee, R.W. and Groffman, P.M. (1997) 'Earthworm abundance and nitrogen mineralization rates along an urban–rural land use gradient', *Soil Biology and Biochemistry*, 29: 427–430.

Stewart, G.H., Ignatieva, M.E., Meurk, C.D., Buckley, H., Horne, B. and Braddick, T. (2009) 'URban Biotopes of Aotearoa New Zealand (URBANZ) (I): composition and diversity of temperate urban lawns in Christchurch', *Urban Ecosystems*, 12: 233–248.

Stewart, I.D. (2011) 'A systematic review and scientific critique of methodology in modern urban heat island literature', *International Journal of Climatology*, 31: 200–217.

Stiles, A. and Scheiner, S.M. (2010) 'A multi-scale analysis of fragmentation effects on remnant plant species richness in Phoenix, Arizona', *Journal of Biogeography*, 37: 1721–1729.

Strauss, B. and Biedermann, R. (2006) 'Urban brownfields as temporary habitats: driving forces for the diversity of phytophagous insects', *Ecography*, 29: 928–940.

Strohbach, M.W., Haase, D. and Kabisch, N. (2009) 'Birds and the city: urban biodiversity, land use, and socioeconomics', *Ecology and Society*, 14: 31.

Stromberg, J.C., Chew, M.K., Nagler, P.L. and Glenn, E.P. (2009) 'Changing perceptions of change: the role of scientists in Tamarix and river management', *Restoration Ecology*, 17: 177–186.

Strubbe, D. and Matthysen, E. (2009) 'Establishment success of invasive ring-necked and monk parakeets in Europe', *Journal of Biogeography*, 36: 2264–2278.

Subburayalu, S. and Sydnor, T.D. (2012) 'Assessing street tree diversity in four Ohio communities using the weighted Simpson index', *Landscape and Urban Planning*, 106: 44–50.

Sudha, P. and Ravindranath, N.H. (2000) 'A study of Bangalore urban forest', *Landscape and Urban Planning*, 47: 47–63.

Sukopp, H. (1998) 'Urban ecology: scientific and practical aspects', in J. Breuste, H. Feldmann and O. Uhlmann (eds) *Urban Ecology*. Berlin: Springer.

Sukopp, H. (2002) 'On the early history of urban ecology in Europe', *Preslia*, 74: 373–393.

Sustek, Z. (1999) 'Light attraction of carabid beetles and their survival in the city centre', *Biologia*, 54: 539–551.

Tait, C.J., Daniels, C.B. and Hill, R.S. (2005) 'Changes in species assemblages within the Adelaide Metropolitan Area, Australia, 1836–2002', *Ecological Applications*, 15: 346–359.

Tan, K.W. (2006) 'A greenway network for Singapore', *Landscape and Urban Planning*, 76: 45–66.

Tansley, A.G. (1935) 'The use and abuse of vegetational concepts and terms', *Ecology*, 16: 284–307.

Taylor, K.G. and Owens, P.N. (2009) 'Sediments in urban river basins: a review of sediment-contaminant dynamics in an environmental system conditioned by human activities', *Journal of Soils and Sediments*, 9: 281–303.

Thames Water (2012) 'Thames Tideway Tunnel'. Online. Available: http://www.thameswater.co.uk/cps/rde/xchg/corp/hs.xsl/10115.htm (accessed 12 August 2012).

Thompson, K. (2006) *No Nettles Required: The Reassuring Truth About Wildlife Gardening*. London: Eden Project Books.

Thompson, K., Austin, K.C., Smith, R.M., Warren, P.H., Angold, P.G. and Gaston, K.J. (2003) 'Urban domestic gardens (I): putting small-scale plant diversity in context', *Journal of Vegetation Science*, 14: 71–78.

Thompson, K., Hodgson, J.G., Smith, R.M., Warren, P.H. and Gaston, K.J. (2004) 'Urban domestic gardens (III): composition and diversity of lawn floras', *Journal of Vegetation Science*, 15: 373–378.

Tian, G.J., Wu, J.G. and Yang, Z.F. (2010) 'Spatial pattern of urban functions in the Beijing metropolitan region', *Habitat International*, 34: 249–255.

Toms, M.P. and Newson, S.E. (2006) 'Volunteer surveys as a means of inferring trends in garden mammal populations', *Mammal Review*, 36: 309–317.

Tóthmérész, B., Mathe, I., Balazs, E. and Magura, T. (2011) 'Responses of carabid beetles to urbanization in Transylvania (Romania)', *Landscape and Urban Planning*, 101: 330–337.

Tratalos, J., Fuller, R.A., Evans, K.L., Davies, R.G., Newson, S.E., Greenwood, J.J.D. and Gaston, K.J. (2007a) 'Bird densities are associated with household densities', *Global Change Biology*, 13: 1685–1695.

Tratalos, J., Fuller, R.A., Warren, P.H., Davies, R.G. and Gaston, K.J. (2007b) 'Urban form, biodiversity potential and ecosystem services', *Landscape and Urban Planning*, 83: 308–317.

Tremblay, M.A. and St Clair, C.C. (2009) 'Factors affecting the permeability of transportation and riparian corridors to the movements of songbirds in an urban landscape', *Journal of Applied Ecology*, 46: 1314–1322.

Tremblay, M.A. and St Clair, C.C. (2011) 'Permeability of a heterogeneous urban landscape to the movements of forest songbirds', *Journal of Applied Ecology*, 48: 679–688.

Troy, A.R., Grove, J.M., O'Neil-Dunne, J.P.M., Pickett, S.T.A. and Cadenasso, M.L. (2007) 'Predicting opportunities for greening and patterns of vegetation on private urban lands', *Environmental Management*, 40: 394–412.

Trudgill, S. (2008) 'A requiem for the British flora? Emotional biogeographies and environmental change', *Area*, 40: 99–107.

Tsai, Y.H. (2005) 'Quantifying urban form: compactness versus "sprawl" ', *Urban Studies*, 42: 141–161.

Turner, K., Lefler, L. and Freedman, B. (2005) 'Plant communities of selected urbanized areas of Halifax, Nova Scotia, Canada', *Landscape and Urban Planning*, 71: 191–206.

Turner, T. (2005) *Garden History: Philosophy and Design 2000 BC–2000 AD*. New York: Spon Press.

Turner, W.R. (2006) 'Interactions among spatial scales constrain species distributions in fragmented urban landscapes', *Ecology and Society*, 11: 6.

Turner, W.R. and Tjorve, E. (2005) 'Scale-dependence in species–area relationships', *Ecography*, 28: 721–730.

Tyrvainen, L., Makinen, K. and Schipperijn, J. (2007) 'Tools for mapping social values of urban woodlands and other green areas', *Landscape and Urban Planning*, 79: 5–19.

Ulfarsson, G.F. and Carruthers, J.I. (2006) 'The cycle of fragmentation and sprawl: a conceptual framework and empirical model', *Environment and Planning B*, 33: 767–788.

UNFPA (United Nations Population Fund) (2007) *Urbanization: A Majority in Cities*. Online. Available: http://www.unfpa.org/pds/urbanization.htm (accessed 26 November 2011).

UN-Habitat (2011) *State of the World's Cities 2010/2011*. Online. Available: http://www.unhabitat.org/content.asp?cid=8051&catid=7&typeid=46 (accessed 22 August 2012).

United Nations Statistics Division (2011) *Definitions*. Online. Available: http://unstats.un.org/unsd/demographic/sconcerns/densurb/densurbmethods.htm#D (accessed 31 January 2011).

United States Soil Conservation Service (1975) *Urban Hydrology for Small Watersheds*. USDA Soil Conservation Service Technical Release No. 55. Washington, DC.

Unlu, I., Farajollahi, A., Healy, S.P., Crepeau, T., Bartlett-Healy, K., Williges, E., Strickman, D., Clark, G.G., Gaugler, R. and Fonseca, D.M. (2011) 'Area-wide management of *Aedes albopictus*: choice of study sites based on geospatial characteristics, socioeconomic factors and mosquito populations', *Pest Management Science*, 67: 965–974.

Uy, P.D. and Nakagoshi, N. (2008) 'Application of land suitability analysis and landscape ecology to urban greenspace planning in Hanoi, Vietnam', *Urban Forestry and Urban Greening*, 7: 25–40.

Valtonen, A., Saarinen, K. and Jantunen, J. (2007) 'Intersection reservations as habitats for meadow butterflies and diurnal moths: guidelines for planning and management', *Landscape and Urban Planning*, 79: 201–209.

Vandersmissen, M.H., Villeneuve, P. and Thériault, M. (2003) 'Analyzing changes in urban form and commuting time', *The Professional Geographer*, 55: 446–463.

Van der Veken, S., Verheyen, K. and Hermy, M. (2004) 'Plant species loss in an urban area (Turnhout, Belgium) from 1880 to 1999 and its environmental determinants', *Flora*, 199: 516–523.

Vergnes, A., Le Viol, I. and Clergeau, P. (2012) 'Green corridors in urban landscapes affect the arthropod communities of domestic gardens', *Biological Conservation*, 145: 171–178.

Vierling, K.T. (2000) 'Source and sink habitats of Red-winged Blackbirds in a rural/suburban landscape', *Ecological Applications*, 10: 1211–1218.

Vimal, R., Mathevet, R. and Thompson, J.D. (2011) 'The changing landscape of ecological networks', *Journal for Nature Conservation*, 20: 49–55.

Virden, R.J. and Walker, G.J. (1999) 'Ethnic/racial and gender variations among meanings given to, and preferences for, the natural environment', *Leisure Sciences*, 21: 219–239.

Vitousek, P.M., D'Antonio, C.M., Loope, L.L., Rejmanek, M. and Westbrooks, R. (1997) 'Introduced species: a significant component of human-caused global change', *New Zealand Journal of Ecology*, 21: 1–16.

Wacker, L., Baudois, O., Eichenberger-Glinz, S. and Schmid, B. (2008) 'Environmental heterogeneity increases complementarity in experimental grassland communities', *Basic and Applied Ecology*, 9: 467–474.

Wackernagel, M. and Rees, W. (1996) *Our Ecological Footprint: Reducing Human Impacts on the Earth*. Philadelphia: New Society Publishers.

Wahlbrink, D. and Zucchi, H. (1994) 'Occurrence and settlement of carabid beetles on an urban railway embankment: a contribution to urban ecology', *Zoologische Jahrbuecher Abteilung fuer Systematik Oekologie und Geographie der Tiere*, 121: 193–201.

Walsh, C.J., Roy, A.H., Feminella, J.W., Cottingham, P.D., Groffman, P.M. and Morgan, R.P. (2005) 'The urban stream syndrome: current knowledge and the search for a cure', *Journal of the North American Benthological Society*, 24: 706–723.

Wandeler, P., Funk, S.M., Largiader, C.R., Gloor, S. and Breitenmoser, U. (2003) 'The city-fox phenomenon: genetic consequences of a recent colonization of urban habitat', *Molecular Ecology*, 12: 647–656.

Wania, A., Kühn, I. and Klotz, S. (2005) 'Plant richness patterns in agricultural and urban landscapes in Central Germany: spatial gradients of species richness', *Landscape and Urban Planning*, 75: 97–110.

Warren, P.S., Harlan, S.L., Boone, C., Lerman, S.B., Shochat, E. and Kinzig, A.P. (2010) 'Urban ecology and human social organisation', in K.J. Gaston (ed.) *Urban Ecology*. Cambridge: Cambridge University Press.

Weckel, M.E., Mack, D., Nagy, C., Christie, R. and Wincorn, A. (2010) 'Using citizen science to map human–coyote interaction in suburban New York, USA', *Journal of Wildlife Management*, 74: 1163–1171.

Weeda, E.J. (2011) 'Maastricht', in J.G. Kelcey and N. Muller (eds) *Plants and Habitats of European Cities*. New York: Springer.

Wenger, S.J., Roy, A.H., Jackson, C.R., Bernhardt, E.S., Carter, T.L., Filoso, S., Gibson, C.A., Hession, W.C., Kaushal, S.S., Martí, E., Meyer, J.L., Palmer, M.A., Paul, M.J., Purcell, A.H., Ramírez, A., Rosemond, A.D., Schofield, K.A., Sudduth, E.B. and Walsh, C.J. (2009)

'Twenty-six key research questions in urban stream ecology: an assessment of the state of the science', *Journal of the North American Benthological Society*, 28: 1080–1098.

Westermann, J.R., von der Lippe, M. and Kowarik, I. (2011) 'Seed traits, landscape and environmental parameters as predictors of species occurrence in fragmented urban railway habitats', *Basic and Applied Ecology*, 12: 29–37.

Wheater, C.P. (2010) 'Walls and paved surfaces: urban complexes with limited water and nutrients', in I. Douglas, D. Goode, M. Houck and R. Wang (eds) *The Routledge Handbook of Urban Ecology*. London: Routledge.

White, E.V. and Gatersleben, B. (2011) 'Greenery on residential buildings: does it affect preferences and perceptions of beauty?', *Journal of Environmental Psychology*, 31: 89–98.

White, R. and Engelen, G. (1993) 'Cellular-automata and fractal urban form: a cellular modelling approach to the evolution of urban land-use patterns', *Environment and Planning A*, 25: 1175–1199.

Wilby, R.L. and Perry, G.L.W. (2006) 'Climate change, biodiversity and the urban environment: a critical review based on London, UK', *Progress in Physical Geography*, 30: 73–98.

Williams, R. (1973) *The Country and the City*. New York: Oxford University Press.

Wimberly, M.C. (2006) 'Species dynamics in disturbed landscapes: when does a shifting habitat mosaic enhance connectivity?', *Landscape Ecology*, 21: 35–46.

With, K.A. and Pavuk, D.M. (2011) 'Habitat area trumps fragmentation effects on arthropods in an experimental landscape system', *Landscape Ecology*, 26: 1035–1048.

Wood, B.C. and Pullin, A.S. (2002) 'Persistence of species in a fragmented urban landscape: the importance of dispersal ability and habitat availability for grassland butterflies', *Biodiversity and Conservation*, 11: 1451–1468.

Wood, W.E. and Yezerinac, S.M. (2006) 'Song sparrow (*Melospiza melodia*) song varies with urban noise', *The Auk*, 123: 650–659.

Wu, J. and Bauer, M.E. (2012) 'Estimating net primary production of turfgrass in an urbansuburban landscape with QuickBird imagery', *Remote Sensing*, 4: 849–866.

Wu, J.G., Jenerette, G.D., Buyantuyev, A. and Redman, C.L. (2011) 'Quantifying spatiotemporal patterns of urbanization: the case of the two fastest growing metropolitan regions in the United States', *Ecological Complexity*, 8: 1–8.

Xie, Y., Yu, M., Bai, Y. and Xing, X. (2006) 'Ecological analysis of an emerging urban landscape pattern – desakota: a case study in Suzhou, China', *Landscape Ecology*, 21: 1297–1309.

Xu, C., Liu, M.S., Zhang, C., An, S.Q., Yu, W. and Chen, J.M. (2007) 'The spatiotemporal dynamics of rapid urban growth in the Nanjing metropolitan region of China', *Landscape Ecology*, 22: 925–937.

Xu, C., Ye, H. and Cao, S. (2011) 'Constructing China's greenways naturally', *Ecological Engineering*, 37: 401–406.

Yang, G.X., Bowling, L.C., Cherkauer, K.A., Pijanowski, B.C. and Niyogi, D. (2010) 'Hydroclimatic response of watersheds to urban intensity: an observational and modelingbased analysis for the White River Basin, Indiana', *Journal of Hydrometeorology*, 11: 122–138.

Yang, J., Yu, Q. and Gong, P. (2008) 'Quantifying air pollution removal by green roofs in Chicago', *Atmospheric Environment*, 42: 7266–7273.

Yasuda, M. and Koike, F. (2006) 'Do golf courses provide a refuge for flora and fauna in Japanese urban landscapes?', *Landscape and Urban Planning*, 75: 58–68.

Yong, Y., Zhang, H., Wang, X. and Schubert, U. (2010) 'Urban land-use zoning based on ecological evaluation for large conurbations in less developed regions: case study in Foshan, China', *Journal of Urban Planning and Development*, 136: 116–124.

York, A.M., Shrestha, M., Boone, C.G., Zhang, S.A., Harrington, J.A., Prebyl, T.J., Swann, A., Agar, M., Antolin, M.F., Nolen, B., Wright, J.B. and Skaggs, R. (2011) 'Land fragmentation under rapid urbanization: a cross-site analysis of Southwestern cities', *Urban Ecosystems*, 14: 429–455.

Young, C.H. and Jarvis, P.J. (2001) 'Assessing the structural heterogeneity of urban areas: an example from the Black Country (UK)', *Urban Ecosystems*, 5: 49–69.

Yuen, B. and Hien, W.N. (2005) 'Resident perceptions and expectations of rooftop gardens in Singapore', *Landscape and Urban Planning*, 73: 263–276.

Yuen, B., Yeh, A., Appold, S.J., Earl, G., Ting, J. and Kwee, L.K. (2006) 'High-rise living in Singapore public housing', *Urban Studies*, 43: 583–600.

Zalasiewicz, J. (2008) *The Earth After Us: What Legacy Will Humans Leave in the Rocks?* Oxford: Oxford University Press.

Zhang, L.Q. and Wang, H.Z. (2006) 'Planning an ecological network of Xiamen Island (China) using landscape metrics and network analysis', *Landscape and Urban Planning*, 78: 449–456.

Zhang, Y., Yang, Z.F. and Yu, X.Y. (2009) 'Evaluation of urban metabolism based on emergysic synthesis: A case study for Beijing (China)', *Ecological Modelling*, 220: 1690–1696.

Zipperer, W.C. (2010) 'The process of natural succession in urban areas', in I. Douglas, D. Goode, M. Houck and R. Wang (eds) *The Routledge Handbook of Urban Ecology*. London: Routledge.

Zipperer, W.C., Foresman, T.W., Sisinni, S.M. and Pouyat, R.V. (1997) 'Urban tree cover: an ecological perspective', *Urban Ecosystems*, 1: 229–247.

Zomlefer, W.B. and Giannasi, D.E. (2005) 'Floristic survey of Castillo de San Marcos National Monument, St. Augustine, Florida', *Castanea*, 70: 222–236.

Zucco, C.A., Oliveira-Santos, L.G.R. and Fernandez, F.A.S. (2011) 'Protect Brazil's land to avert disasters', *Nature*, 470: 335.

Index

Note: page numbers in *italic* type refer to Figures; those in **bold** refer to Tables.